Valuable Broadleaved Forests in Europe

VOLUME 22

Valuable Broadleaved Forests in Europe

European Forest Institute Research Report 22

Edited by

Heinrich Spiecker
Sebastian Hein
Kaisu Makkonen-Spiecker
Michael Thies

EUROPEAN FOREST INSTITUTE

TUTA SUB AEGIDE PALLAS 1683

BRILL

LEIDEN • BOSTON
2009

The views expressed in this book are those of the authors and do not necessarily correspond to those of the European Forest Institute.

This book is printed on acid-free paper.

Library of Congress Cataloging-in-Publication Data

Valuable broadleaved forests in Europe / by Heinrich Spiecker ... [et al.].
p. cm. — (European Forest Institute research report ; 22)
ISBN 978-90-04-16795-7 (hardback : alk. paper) 1. Forests and forestry—Europe.
2. Timber—Europe. I. Spiecker, Heinrich. II. Title. Series.

SD177.V35 2009
634.9'5—dc22

2009001095

ISSN 1238-8785
ISBN 978 90 04 16795 7

PRINTED IN THE NETHERLANDS

Table of Contents

Executive Summary

Kaisu Makkonen-Spiecker
Institute for Forest Growth, University of Freiburg, Germany

Which trees are valuable?

Global demands for sustainable forest management, elevated public concerns for the environment and concerns for the biodiversity of natural systems, as well as the changes in the structure of the global timber market, demand innovative responses from forest managers. Forest managers have recognised for a long time the importance of valuable broadleaved species in increasing the heterogeneity and diversity in European forests. However, their use as timber has gained new appreciation in small-scale wood industries, house building, and wood crafting. Thus, the question of which species are the valuable ones is subjective and is often a matter of personal perspective derived from economic and social values tempered by space and time.

Broadleaved species are often identified as being noble or valuable for a variety of reasons: their high timber value, the beauty of their wood, their fast growth, or their unique site requirements. Certain broadleaved species can also be valuable because of their rarity, beauty or cultural significance. When forest experts were asked to list the most economically important broadleaved species whose importance seems likely to continue to increase in the future, they named common ash (*Fraxinus excelsior* L.), sycamore (*Acer pseudoplatanus* L.) and wild cherry (*Prunus avium* L.) (see Chapter 2.2). The reasons for their prediction is based on (1) the increasing proportions of these three species in the total forested area, (2) higher proportions of these species in younger age classes and (3) an increase in using these species for high valued veneer logs. Accordingly, these three species, which are referred to as valuable broadleaved species, are the main subjects of this book.

Although common ash, sycamore, and wild cherry are the main focus of the book, aspects of certain species of other genera are also included, e.g., *Alnus, Carpinus, Castanea, Juglans, Malus, Pyrus, Sorbus, Tilia,* and *Ulmus.* Beech and oak may produce valuable timber as well and are common in many parts of Europe. However, as these two species are dealt with in many publications, they are not included in this book.

It is important to note that the proportion (area and volume) of valuable broadleaves rarely exceeds 5% in European forests considered in aggregate. However, the proportion can increase up to 25% of the total forest area when smaller areas are considered such as those in local forest enterprises and specific forested sites like floodplain forests or forest plantations (see Appendix B).

A heterogeneous group of species with different silvicultural management regimes

The species that compose the valuable broadleaves group have diverse and often demanding silvicultural requirements. Valuable broadleaves usually grow best on fertile sites (chapter 2.1), but over the centuries, much of this fertile land was cultivated, and consequently human activity substantially reduced the availability of suitable sites for these species. Apart from lime (*Tilia* spp.) and hornbeam (*Carpinus betulus* L.), most valuable broadleaves are light-demanding species which can have difficulty establishing themselves in closed canopy forest conditions where canopy disturbance is infrequent. Moreover, most of the studied valuable broadleaved species are not very tall at maturity. For example only common ash can exceed the maximum height of beech (40 m) while sycamore and sweet chestnut (*Castanea sativa* Mill.) may reach the height of oaks (around 35 m). Thus valuable broadleaves may find themselves at a serious disadvantage in competing for light.

Except for lime, sycamore and sweet chestnut, valuable broadleaved trees are rather short-lived (70–150 years). Nonetheless, many species have the potential to produce high quality timber at a relatively young age, even as young as 60 years or less.

General silvicultural guidelines are difficult to develop for the valuable broadleaves, because each species reacts differently to management regimes and site conditions. However, one common feature is that on most sites they usually require more frequent silvicultural interventions for survival and adequate development than other species, such as beech.

We have examined the suitability of each of the traditional main continuous-cover silvicultural systems (group selection, shelterwood, and coppice-with-standards) to the species characteristics of the valuable broadleaves. We conclude that, apart from the limes and hornbeam, the most appropriate silvicultural system for the valuable broadleaves is group selection using groups of about 0.5 ha, or on appropriate sites, shelterwood or coppice-with-standards systems. The authors embrace the idea that silvicultural systems should be considered as a continuum from clear-cutting to single tree selection, with each point along the continuum representing an opportunity to modify the system to meet different combinations of structural objectives for the stand. Under this future scenario for applying a continuous-cover system, the authors encourage considering and emphasizing those stand elements of that contribute to heterogeneity which is often supplied by the presence of the valuable broadleaves. In the future, wherever valuable broadleaves exist, silvicultural strategies should aim to encourage their regeneration, primarily by adopting a more flexible approach to managing gaps in the overstorey of the stand. However this may entail more frequent and therefore more expensive silvicultural interventions than in pure species stands, in order to ensure the survival and the desired diameter growth of the less competitive valuable broadleaved species.

One of the most serious impediments to implementing a silviculture that promotes valuable broadleaves, and ecologically sensitive management of forests in general, is cost. There seems to be an increasing divergence between economic and ecological objectives of forestry and one of the most pressing needs today is to develop new systems or to adapt current systems that can bring the two objectives together.

However, this may mean that using new and innovative ways of managing forests for multiple objectives will probably involve harvesting at different scales within the same species/site unit thereby requiring more frequent interventions than is currently practiced. But a change to more frequent stand entries may give valuable broadleaves a greater chance to survive and grow in European woodlands and thereby contribute importantly to timber production, diversity and social values.

Importance of high genetic quality for growing valuable broadleaves

Genetic quality is highly important for valuable timber production (chapter 3.1). The flowering biology, natural range, and genetic studies and provenance experiments for sycamore, common ash, and wild cherry are addressed in this section. Also, producing high quality seeds in seed orchards and vegetative propagation of superior clones for further improvement are described. Finally, conservation activities for a range of valuable broadleaved species, legal issues, and recommendations for using reproductive material are outlined.

While several value-relevant tree traits such as diameter growth can be modified considerably by spacing and competition, and thus can be controlled by silviculture, phenological traits such as leaf flushing and bud set are under strong genetic control and can only slightly be modified by changes in the environment. Moreover growing stands with economically valuable phenotypes is only possible if both genotype and environment components are favourable.

The main purpose of silviculture and tree breeding is to develop phenotypes that serve human needs. Silviculture mainly influences the environment surrounding the trees. Although, to a certain extent, thinning can alter the genetic composition of the stand, its ultimate effect depends on the heritability characteristics of the species. On the other hand, tree breeding influences the genotypes of the trees by purposeful selection of individual trees, controlled crosses and choice of propagation methods. Provenance, progeny and clonal trials and tests are tools used to determine existing variation among tree populations and for improving individual tree quality by identifying well performing genetic collections. Tests are established under different environmental conditions to get information about the genetic control of phenological traits and the general adaptability of the genetic units to environmental changes. This information cannot be gained from marker studies alone. Important characteristics of trees that are amenable to adaptation include phenological, survival, vitality and health traits. These traits enable trees to adapt to changes over space and time. In addition, tree improvement through genetic control can modify growth, quality, and wood characteristics. Provenance experiments with valuable broadleaved species are comparatively new. Establishing provenance experiments with these species requires approaches different than those of major tree species composing extensive stands. For example, the approach of collecting a limited number of single tree progenies per population is a compromise to obtain information about provenances as well as progenies. Pragmatically, a forest owner can introduce improved material into the forest when the stand is regenerated without significantly increasing the total cost of regeneration because the recommended number of planted trees per hectare would be small. Thus planting genetically improved material to regenerate the forest is a very effective way for the forest manager to reduce regeneration costs and increase his/her future economic return.

The authors conclude this section by observing that the awareness of what are now called "valuable broadleaved species" in forests has changed considerably historically and that that change has genetic implications. A lack of interest in managing these species in the past has had a major influence on their gene populations. Local extinction, fragmentation, pollination by cultivars and unregulated seed transfer changed the original pattern of variation in forest stands considerably and resulted in unpredictable gene pools. Therefore variation in seed and stock sources should be taken into account by forest owners and forest managers when selecting reproductive material for planting these species. Fortunately, the efforts of genetic conservation and tree breeding for these species during the last 50 years can help to overcome some of the obvious mistakes created by unregulated human interferences during previous centuries, and at the same time help improve the stability, health and productivity of these increasingly economically interesting species.

Pests and diseases

Pests, diseases and other disorders can pose life-threatening problems for valuable broadleaved tree species and effectively prevent their economic success on certain sites or under certain circumstances (chapter 3.2). Potentially damaging factors must be identified before making a decision to plant or cultivate a certain tree species in forests or in plantations. In many cases, the possibility of controlling diseases and pests is very limited, especially when considering the use of chemical control methods, which have been the focus of much research in the past. However, recent knowledge of the predisposition of trees to pests and diseases, genetic qualities of the host tree, and factors in life cycles of diseases and pests often provides information necessary to guide the selection of a tree species that is suitable for a given site and thereby reduce the risk of insect and disease attacks.

Some important pest organisms not native to Europe have either been recently introduced or could pose a real threat to wood production if they were introduced. Such species usually possess high damage potential since they have no or few natural enemies in Europe. For example, *Sirococcus* canker of walnut (*Juglans* spp.) and Asian long horned beetle (*Anoplophora glabripennis*) are two exotic species that, although not present now, are looming on the horizon. However, it is difficult to predict if and when these species will emerge in Europe and the extent of the damage they will cause to valuable broadleaved tree species if they occur.

Crown architecture, morphological variations

The chronological sequence of crown development of wild cherry, common ash and sycamore as well as various morphological adaptations of the crown to environmental factors are described in Chapter 4.1. In order to document how plant morphology changes over time, several individuals at various stages of development were studied from the time of germination to death and then the data collected from these observations were analysed by crown development characteristics. Growth dynamics of the stems of old trees were determined *a posteriori* by identifying comparable morphological markers present in the primary meristems of both old and young trees.

From germination to mature stages, the tree develops a main stem which bears branches that in turn become progressively more ramified. The architecture of the mature tree presents a very precise organisation and is characterised by the nature and relative position of discernible morphological categories of axes.

For example, young wild cherry trees exhibit a cone-shape crown and possess a strongly hierarchical and simple structure composed of a main stem bearing branches, branchlets and short axes. As the trees mature, the crown develops into an irregular oval shape, and in old age the axes bend down.

In ash, a young tree presents a cone-shape crown organised around a single stem (excurrent). As the tree becomes older the crown becomes spherical and develops into a more complex structure made up of numerous main strong and vertical limbs (deliquescent). This crown transformation is the result of an architectural metamorphosis corresponding to the duplication of the architectural unit during tree development.

Flowering characteristics vary among the valuable broadleaves. Monopodial[1] species such as wild cherry and ash exhibit a hierarchical architecture compared to sympodial[2] species such as sycamore and walnut. However, unlike wild cherry and common ash, flowering in sycamore is terminal and the initial monopodial development of the young tree becomes sympodial in the adult stage. In the mature tree terminal flowering of the main stem involves the development of relay axes at its extremity. At first, one dominant relay axis normally develops in continuity with the main stem thus allowing it to keep a rectilinear aspect. Later the dominance of one of the relay axes disappears and a fork is produced at the top of the main stem. The crown is then built up by a succession of forks.

The crown architecture is a consequence of interspecific variations in the developmental pattern of axes. However, variations in branching pattern may also result in different adaptive forms because of environmental conditions. Stem development may be modified by factors such as light availability and population density. Thus for a given species architectural features may be strongly affected by environmental factors. For example, for all genotypes environmental conditions mostly modified the branch length to stem length ratio and the time and height at which the main crown limbs appear. This means that the length of the butt log will be shorter in full light (open-grown trees) or in conditions of low fertility, whereas growth in a forest gap or with lateral shelter (shade) will result in the development of a longer butt log and thin horizontal branches. The authors use results of studies of these of genotype and phenotype characteristics to define the main qualitative architectural features of some of the valuable broadleaved species and their intra- and interspecific variations. Knowing the architectural diversity and plasticity of species provides

1 growing upward with a single main stem or axis that produces leaves and flowers
2 having or involving the formation of an apparent main axis from successive secondary axes

valuable information which can be used to identify the physiological status of trees, to improve forest and plantation management with respect to timber quality, or to conduct breeding and genetic improvement programmes.

Pruning dynamics and diameter control

The presence of branches and knots greatly affect wood quality. Chapter 4.2 describes the natural pruning dynamics of common ash, sycamore and wild cherry. Moreover, the chapter includes a model to predict the length of the branch-free bole, which is important for controlling wood quality. The model includes parameters for total height of the tree, diameter at breast height, age of the tree and whether the tree is forked or not. Also, stem form functions are used to predict the diameter of the knotty core given site quality and competition. Then decision rules are established to control the length of the branch free bole and the width of the knotty core. Using the results of this model, a two-phase system for controlling natural pruning dynamics is presented.

Models of branching dynamics presented in this chapter give forest managers a tool which can be used to improve wood quality. The main variables affecting wood quality are the length of the branch free bole and the width of the knotty core. But, there are other crucial decisions a forest manager has to make which are closely linked to branching: setting the target diameter and the production time and therefore, by extension, stem radial growth. Thus decisions on maintaining a certain stand density and the number of crop trees per hectare are very important for producing branch-free wood volume. Additional considerations include species-specific differences in pruning dynamics, crown width development and stem form. The model presented in this chapter and the crown width model in Chapter 4.3 can be used to quantify these individual tree and stand relationships. For example, when the length of the branch-free bole decreases there will be an increase in mean radial increment, which means there has to be a smaller number of crop trees per hectare if target diameter at breast height is large. Accordingly, the total volume of externally branch-free wood per ha decreases as a result. On the other hand, for a given mean radial increment, the total volume of externally branch-free wood per hectare increases with increasing site quality since trees grow faster in height on good sites compared to poor sites, other factors held constant.

Differences in productivity related to natural pruning dynamics, number of crop trees per hectare and stem form among species are reflected in a comparison between common ash and sycamore growing on the same site: the per hectare volume of branch-free wood in common ash is higher than that of sycamore. Moreover, two other differences become apparent when one considers in more detail the volume-related proportion of the branch-free wood with a knotty core exceeding one-third of the diameter of the trunk: (1) as mean radial increment increases the portion of this volume increases as well; and (2) as site quality increases the proportion of the externally branch-free wood with a knotty core exceeding one-third of the diameter of the trunk decreases at a fixed mean radial increment. This helps explain why the per-hectare production of externally branch-free wood of common ash is higher than that of sycamore, all else being equal.

When producing high quality timber, the implications of diameter growth on quality and especially on natural pruning dynamics must not be neglected (Chapter 4.3). A sudden heavy release of trees will result in an acceleration of diameter growth, the development of epicormic branches, thick branches, and a large knotty core. If these trees subsequently experience strong inter-tree competition, large branches may die. These branches reduce wood quality and may be the gateway for fungi causing wood decay. Besides branchiness, large radial increments normally result in poor mechanical properties of the wood. But in the case of common ash, to the contrary, large growth rings may actually improve the mechanical properties of the wood.

To better reflect the current variety of targeted harvesting diameters of mature trees, Chapter 4.3 discusses diameter objectives referred to in the literature and forest management practices to achieve them. Based on these objectives, methods are derived for forest management with special emphasis on controlling diameter growth. Forest managers can control diameter growth by deciding (1) on a target diameter, (2) a production time, and (3) then by calculating average radial increment. All decision tools presented in Chapter 4.3 are based on deciding on a desired target diameter or a desired mean radial increment. For changes in diameter growth, deviations from values given in the chapter can be made. In conclusion, the authors recommend starting thinning early and maintaining a short interval between successive thinnings to produce high quality timber.

Final cutting systems

Very little information on final cutting systems of valuable broadleaves is available. Nevertheless, Chapter 4.4 deals with the size of final cutting units as well as with the timing and the duration of the final cutting.

Stand history and composition of valuable broadleaves, especially their position and behaviour with respect to the major tree species in the stand, strongly influence the cutting system. Common ash, sycamore and wild cherry usually occur on small areas, intimately or group-mixed with more common hardwood species, such as beech and oak, which have different silvicultural characteristics. But, sometimes valuable broadleaves appear in associations in which they are dominant. Recently these species have been planted and treated in groups or in small monocultures.

Since common ash, sycamore and wild cherry share a number of common silvicultural characteristics, the final cutting systems for these species are similar. But differences do exist, especially concerning management objectives and management systems among different forest regions. When considering the size and duration of final cutting systems of valuable broadleaves, three typical systems are used: (1) single tree cutting; (2) group cutting; and (3) small stand cutting. Typical clear-felling, shelterwood and selection cutting are not normally applied to these species. Since valuable tree species normally occur in intimately mixed (either with oak or beech) high forests, final cuttings often occur on a single tree level. However, a group cutting system should be applied in beech and oak stands that are group-mixed with the valuable broadleaves, in stands where these latter species are dominant and group-mixed, and in pure plantations of these species. The size of the groups usually ranges from 0.1 to 0.5 ha. The duration of the final cuttings depends on several factors, although generally periods of 10- to 20-years are recommended. Recently established small plantations of common ash, sycamore and wild cherry will normally lead to cutting systems that are applied on a stand level. The cutting area will vary around 1 ha, and all trees will normally be cut at the same time. Such plantations and cutting systems are attractive for private forest owners, but can also be recommended for public forests.

A fixed rotation period for the final cutting system should be avoided. Most authors recommend a rotation period of 60- to 80-years for common ash, sycamore and wild cherry. The actual length of the period is determined by several factors, of which site characteristics, growth potentials, production goals and timber quality are the major elements. Of these, site quality strongly influences growth pattern and consequently rotation period. In order to produce valuable timber target diameters should be within a range of 50–60 cm. But to avoid a greater risk of timber discoloration and deterioration after the age of 60- to 80-years, long rotation periods and very large diameters should be avoided. However there are exceptions: rotation length for Norway maple (*Acer platanoides* L.) might reach 100- to 120-years, while for wild service tree (*Sorbus torminalis* L.) a rotation of 120- to 150-years would be acceptable.

Wood properties and wood processing

Common ash and sycamore are very valuable because of their excellent properties and suitability for use in fine furniture and other indoor uses (Chapter 4.5). However, this valuation applies only to the best wood quality, and requires faultless properties and log diameters of at least 30 cm. Inferior quality wood – devalued by knots, unfavourable stem form, and smaller dimension– barely covers the cost of production and are in less demand. Unfortunately, the proportion of better quality wood has been rather low, and needs to be increased in the future. Therefore, silvicultural practices such as heavy crown thinning that promote superior quality and larger dimensions in a relatively short time are to be favoured.

Valuable broadleaves enriching the landscape

Valuable broadleaved trees often are part of the landscape in many areas of Europe; in forests, in the countryside, and in cities (Chapter 5.1). The principles underlying forest landscape aesthetics apply equally to valuable broadleaved trees as to other trees. Across Europe there is a range of different "human-based forest cultures" which affect how people relate to forests and trees. Therefore any social consideration of forests and landscapes has to be viewed in this context. Valuable broadleaved trees contribute to the forest landscape by the variety of their colours and textures as they change through the seasons. Interestingly, common ash and sycamore are used infrequently in urban landscapes, while wild cherry or special ornamental varieties of cherry are popular. When these species are also grown for their timber volume and quality, the same management principles should be applied as for other forest types. Plantation layout, shape and scale of woodlands and compartments, the geometry of rows, stand structure and the effects of final cutting all need to be considered as part of landscape planning.

Forest landscape aesthetics is a well-developed discipline which can provide practical advice to forest managers who may wish to engage in intensive silviculture in order to maximise timber quality and volume while avoiding intrusive effects. The presented set of guiding design principles should be interpreted for each specific landscape in which the trees are planted. Their application should respect inherent landscape diversity present across Europe and the regional forest cultures embedded in that landscape. Since the approach to managing valuable broadleaves to enrich the landscape suggested in this chapter has yet to be tested on the ground, there is great opportunity for research into the potential visual impacts of different landscape designs that feature valuable broadleaves. This research could use virtual visualisation technologies to test landscape design alternatives and public reaction to them within different European countries where valuable broadleaved trees are likely to be grown.

Valuable broadleaved trees and nature conservation

Stands containing valuable broadleaved tree species are productive and sustainable with respect to abiotic natural resources. Production forests of these species can be supported on many sites. Concerns related to nature conservation originate from the genetic selection of forest trees, the practice of 'improving' low fertility sites that in turn contributes to the process destroying the corresponding biocenosis, and from the practice of thinning and harvesting the stands before the trees can reach large dimensions. Large dimension trees are important to ecological processes because they provide essential habitats for species typical of old-forest life phases. Because of these concerns several recommendations are made in Chapter 5.2.

(1) Present stands of even-aged, uniformly-sized valuable broadleaved forests can be changed through silvicultural measures like selective thinning thereby resulting in uneven, mixed forest structures. This implies avoiding establishing or maintaining stands consisting only of valuable broadleaved trees but instead, mixing them with shade tolerant trees such as beech (*Fagus sylvatica* L.) or fir (*Abies alba* Mill.),

(2) Extreme low fertility sites must not be meliorated. Tree composition on such sites must not be altered to produce stands of valuable broadleaved tree species. Extreme sites and their biocenosis must be preserved even if they are unproductive.

(3) Due to the artificial genetic selection of trees in production forests, a sufficient number of large forest reserves must be created to preserve the site-

adapted genetic diversity of the tree species. Only forest reserves have the potential to provide terminal and decaying phases of old-growth forests with temporal and structural niches for those species that are absent in production forests, e.g., species inhabiting large-diameter decaying wood. Afforestation of sites with valuable biocenosis (e.g., oligotrophic grasslands) must be avoided.

(4) Measures to avoid environmental destruction, such as those outlined in environmental impact assessments, should increasingly take natural succession into consideration as opposed to afforestation.

Valuable broadleaves as an object of public perception and policy

Chapter 5.3 examines the importance of tree species in the public perception of forests. The chapter gives an overview of theories explaining forest preferences and describes the crucial role of valuable broadleaves in culture and politics. Furthermore, the chapter investigates their relevance in people's minds by reviewing surveys undertaken in Europe.

Assumed preferences of forest visitors are frequently used as a motivation for broadleaved-based forest management. The crucial question *"which species are desired?"* is usually answered unanimously by the experts by stating that *"broadleaved trees are vital elements in multiple use forestry both in rural areas and in urban settings"*.

However, it is important to note that broadleaf mixtures are evaluated by their relative share in the region rather than by absolute cover in specific stands. Furthermore, opinion polls show that preferences for traditional broadleaved forest regions and those dominated by conifers differ between ordinary people and small-scale forest owners. Several studies show an increasing degree of public support for non-coniferous trees while small-scale forest owners prefer conifers. Forest management journals, silvicultural guidelines, forest management plans and political programmes such as forest acts, environmental laws or forestry promotion schemes are good examples of an increasing public awareness about broadleaved trees.

Moreover, trends in opinion polls suggest a growing preference for broadleaved trees among forest visitors. Across Europe, public preferences concerning forest types have been surveyed on local, regional and national levels. In Germany, surveys were compiled for the first time in the early 1970s, and the most recent compilation was completed at the end of the 1990s. The results of this time series show that the preferences in the four most widely applied categories – *broadleaved forest, conifer forests, mixed forests* and *no opinion* – change significantly over time. The preferences for *mixed forests* are stable, whereas the *broadleaved forests* have become more popular at the expense of conifer forests. However, a large proportion of people express *no opinion* on this issue; about 20% of the respondents opted for this category in the surveys where it is used. If no comparable category is offered, people tend to mark *mixed forests* in their questionnaires if they are undecided.

Studies from several different European countries indicate that species is not the most important characteristic for non-professionals evaluating trees and forests. The acceptability of forests and forest management is more likely to be influenced by the size or structure of woodlands, damage, and visible results of management actions.

Contribution of valuable broadleaves to sustainability of European forests

Chapter 6 concludes that valuable broadleaves, as a species group, offer options for increasing ecological, economic and social values and likely contribute importantly to the sustainability of forestry in Europe and other parts of the world. They may increase the production of high quality timber while maintaining and improving environmental services. Thus, valuable broadleaves may help balance timber and non-timber production, while maintaining economic attractiveness of forests. The high diversity of sites, ownerships, economic and socio-cultural conditions in Europe may require different management strategies that are adapted to local needs. New tools for inventories to objectively monitor the development of wood quality, as well as indicators for value-relevant wood properties need to be developed.

Acknowledgements

This book is a joint product of twenty international scientists as authors, supported by the European Forest Institute and coordinated by the Institute for Forest Growth at the University of Freiburg, Germany.

We would like to thank Ms Nancy Hofer for the language review and her support in the editing process, as well as Ms Anneli Winter for her assistance. Our special thanks are directed to Ms Minna Korhonen at EFI for managing the final steps in publishing this book.

We thank the reviewers for their valuable comments and the Scientific Advisory Board of the European Forest Institute for their evaluation.

List of reviewers:

J.-Ch. Bastien, France
J. Bauhus, Germany
T. Cowell, United Kingdom
S. Fink, Germany
P. Gemmel, Sweden
P. Glück, Austria
E. Hochbichler, Austria
O. Holdenrieder, Switzerland
J. Hynynen, Finland
G. Kerr, United Kingdom
F. Kienast, Switzerland
C. Konijnendijk, Denmark
M. Krott, Germany
N.-V. Nicolescu, Romania
A. Oosterbaan, Netherlands
J. Parviainen, Finland
E. Rojas-Briales, Spain
A. Roloff, Germany
H. Schröter, Germany
J.-P. Skovsgaard, Denmark
Z. Somogyi, Hungary
M. Suda, Germany
M. Tomé, Portugal
S. Wagner, Germany
E. Verkasalo, Finland

1. Introduction

1.1 Increasing Interest in Valuable Broadleaved Tree Species

Heinrich Spiecker
Institute for Forest Growth, University of Freiburg, Germany

Valuable broadleaved tree species, such as common ash (*Fraxinus excelsior*), sycamore (*Acer pseudoplatanus*), wild cherry (*Prunus avium*), walnut (*Juglans regia*, *J. nigra* and hybrids), wild service tree (*Sorbus torminalis*), black alder (*Alnus glutinosa*) and lime (*Tilia cordata*) are an increasingly important element of forests in Europe. The natural range of these valuable broadleaved species expands from the boreal zone in the north, to the Mediterranean zone in the south, from the continental zone in the east to the Atlantic zone in the west.

Valuable broadleaved tree species have been neglected over time and to a large extend, removed in former times from their natural sites during the phase of forest devastation and afforestation with other species. The percentage of high forest consisting of coniferous species such as Norway spruce and Scots pine increased continuously, while forest consisting of broadleaved species decreased. Especially the area of broadleaved forest consisting of coppice forests, and coppice with standard forests decreased on many sites. Today's growth rates and growing stocks are high, but at the cost of a shift to non-site-adapted tree species, an increasing average stand age, and a reduction in resistance to storm, snow, ice, droughts, insects, and fungi. The current composition of European forests is the result of centuries of vast human activities. They led to a shift to occasionally non-site-adapted tree species and a reduction in tree species diversity. The shortage of wood in European forests, in addition, favored fast growing species and less demanding species and has effected the composition of tree species substantially (Spiecker et al. 2004). Finally, little or no knowledge on the management of valuable broadleaved tree species reduced the willingness to cultivate them, leading to a low supply of valuable wood.

Valuable broadleaved species are often light demanding, grow best on highly productive sites and are often rather short-lived. They eventually need intensive release and different scales and patterns of canopy disturbance in the phase of regeneration. In most European countries, they typically make up less than 5% of the forest cover and produce less than 5% of the total timber volume. They usually grow individually or in small groups in a mixture with other broadleaved trees and conifers. Mixed with other tree species, they generally need intensive release and in the phase of regeneration, they need high intensity of canopy disturbance. New ecological aspects get apparent, as for example the climate change, which led to an increase in air temperature during the last 100 years on many sites. It is predicted that air temperature will globally increase even more in the future. Favoring site-adapted tree species and increasing tree species diversity may strengthen forest resilience in a changing environment. Valuable broadleaved species may help to improve resilience of forests and reduce ecological risks. Growing valuable timber contributes to carbon sequestration as well. In addition, valuable broadleaved tree species may increase diversity of habitats. Valuable broadleaved species can be grown successfully as part of an existing forest management regime, in private and public forests and in a mixture with both broadleaved and coniferous trees on forest land, after afforestation of farmland, in orchards, along roads or in agro-forestry systems.

As urban forests are becoming increasingly important for people living in cities, and land traditionally managed for forestry is under pressure to change, new ways of managing forests need to be explored. Non-wood products, some of them such as apiculture being traditional, ones, could be a parallel source of livelihood in combination with growing valuable broadleaved tree species. On the other hand, biomass production in short rotation combined with valuable wood production may supply urgently needed renewable resources. Using the ideas of agro-forestry in combination with the management of valuable broadleaved trees may create new and innovative management regimes. New ways of multiple land uses may arise as better habitats offer opportunities for raising livestock, and may also generate jobs and new business ideas, ultimately offering people in rural regions new options for making a living. At the same time, the interest in ecosystem services of forests increased. Today's society in Europe is asking for sustainable multi-purpose forestry emphasizing biodiversity. Values and perceptions of people are changing. Commodity values have decreased, while ecological values such as habitats, water quality, and climate protection have increased. Recreational values of forests have increased as well. Environmental values and ecosystem services of forests and landscapes are emphasized. These changes led to a preference of close to nature forest management which encourages an increase in the proportion of valuable broadleaved tree species. Tree species such as wild cherry may add to the beauty of the landscape through its flowering in spring. They may as well provide pollen for bees and fruits for birds and other animals in summer. Only recently, the public awareness of the need for renewable resources such as timber became evident again. There is no legal obligation to change tree species composition towards more minority tree species. However, it is acknowledged that water quality, soil fertility, genetic diversity, and landscape beauty are of high relevance for society, and a balance between private and public interest has to be found. Forest ecosystems should be stable and elastic and they have to serve multiple purposes. They should provide protection, recreation, as well as sustainable production of valuable timber. A substantial proportion of various site-adapted tree species should be accomplished, which means an increase in broadleaved species including valuable broadleaved species on many sites. In Germany for example, different states offer varying ways for federal subsidies, such as subsidy for regeneration of broadleaved and mixed forests.

While labour cost has increased considerably, the value of wood decreased in the last 50 years. Only recently, timber prices rose again. Even though the total productivity per man hour increased during this time, the net income from forestry decreased. Due to the economic changes new ways of management have to be found. In addition, a globalization of markets has to be acknowledged. The recent expansion of European Union led to an increase of the forest area by 53 million ha in 1995, by 24 million ha in 2004, and a further increase by 10 million ha in 2007. As a consequence of these changes costs have to be reduced in various countries, labour productivity has to be raised, wood quality has to be improved, forest stability and resilience has to be enforced, and the ecological services of forests have to be enhanced. Forests should be managed in a highly efficient way by restricting interventions to actions that have a direct impact on the value production. High quality timber from valuable broadleaved species may contribute to higher revenue without increasing costs. The improvement of the economic conditions primarily depends on the quality of wood. Low quality timber of broadleaves produces less value, eventually not even covering the harvesting costs, whereas high quality timber is many times more valuable. Stem dimension and branchiness are key indicators for timber quality. Even though the price of high quality timber may depend on fashions – some species are preferred at some time while other species are preferred in other times – high quality timber from broadleaved trees was highly valued since many years and presumably the increasing need for renewable resources may further increase the value of such timber in the future. The import of valuable wood from the tropics is in discussion while the need for veneer logs and valuable wood for furniture and panels is continuing. Valuable broadleaved tree species may produce economic values by producing high quality timber and this within a relatively short time.

There are new challenges today. Besides traditional economic criteria, such as net present value, cash flow, and risks, new aims become more important, such as flexibility to adapt to new ecological, economic, social and political constrains. On sites naturally dominated by broadleaves, coniferous forests often are less favourable from the ecological point of view, while economically, conifer forests may still be preferable on many sites. As society is not satisfied with the existing forests, forest conversion has to be considered. To provide a solid base for analyzing the costs and benefits of conversion of forests towards other types of forests with respect to species composition and stand structure, various aspects need to be taken into account (Spiecker et al. 2004). There is no one single optimal forest! The best option depends on site conditions, the state of the forest, economic conditions, such as forest size and location, and the aims of the forest owner. Aims are changing fast when compared to the slow changes of forests. So, forests have to be formed which can be adjusted to changing aims more easily. Valuable broadleaved tree species offer options for increasing ecological, economic and social values and therefore contribute to multi-purpose forestry.

As our value system is changing, the aims of forest management have to adapt to them. A widened scope and new ways of forest management are required. As site conditions, ownership and cultural, economic and social conditions vary at short distances in Europe, also forest management differs locally. Forest managers must find new ways to scope with uncertainty and risk using adaptive and flexible strategies. Diversity in cultivation conditions requires strategies adapted to local and regional needs. Pursuing sustainable forestry will ensure higher resilience of forests, which in turn will increase the economic and social benefits of forests and reduce the risks. Various forest types may form a diverse landscape providing the ecological values needed today and possibly in the future. As valuable broadleaved species contribute to the heterogeneity and diversity of forests and landscapes and as well have the potential to produce high quality timber within a relatively short time, these species are of ecological and economic interest.

Reference

Spiecker, H., Hansen, J., Klimo, E., Skovsgaard, J.P., Sterba, H. and von Teuffel, K. 2004. Norway spruce conversion: Options and Consequences. European Forest Institute Research Report 18. Brill Academic Publishers, Leiden, Boston, 269 p.

1.2 Aim of the Book

Heinrich Spiecker
Institute for Forest Growth, University of Freiburg, Germany

Ecological and economic considerations recently increased the interest in growing valuable broadleaved tree species. Even so the demand for valuable timber is growing, and there is a notable interest among forest owners and farmers to grow valuable broad leaved tree species, the current level of knowledge about these species is insufficient. More information on how to grow valuable broadleaved species to obtain high-quality wood and more research on new options for forest management is needed. The book covers various relevant aspects of growing valuable broadleaved trees, and therefore an interdisciplinary approach is applied. The disciplines are represented by a consortium of experts and professionals in the different disciplines of forest sciences and related areas. They describe the state of the art in their research fields.

The main objective of the book is to increase the knowledge of growing valuable broadleaved tree species with the intent to facilitate growing highly valuable wood in a short production time with low investment of work, energy and capital. New options for management of trees on small and large holdings, along roads, in the field, in private and public forests of various densities and in mixtures with both broadleaved and coniferous trees are considered. Its emphasis lies on developing new management regimes, aiming to effectively produce high-quality hardwood from valuable broadleaved trees with methods adjusted to ownership, tree species, site conditions and type of land and region.

At the same time, new dimensions of growing valuable broadleaved trees will be investigated, i.e. new ideas where and how to grow these species will be developed, while taking the aims and restrictions of the production of non-wood products into consideration. Such products are biodiversity, nature conservation, habitat values, landscape and recreational values, and products that can be collected or gathered as by-products parallel to the main product wood. Non-wood products are to be integrated with the wood production and management objectives.

The book includes biological, technical, as well as social aspects. It lists the tree species covered in this book and their geographic distribution. Species specific site requirements are considered. It highlights the importance of genetics and tree breeding. Diseases, disorders and pests of various tree species are extensively discussed in a separate chapter. Furthermore, the book deals with crown architecture of valuable broadleaved tree species. Growth models are presented and special emphasis is given to value generating aspects such as controlling diameter growth and natural as well as artificial pruning. Systems for determining the final cut are developed. Value relevant wood properties and wood processing are shortly demonstrated. The visual impact of valuable broadleaved trees in the landscape is analyzed and the contributions of growing valuable broadleaved trees to nature conservation are considered. Finally, the public perception of valuable broadleaved trees and policy implications are highlighted.

By the interdisciplinary approach new concepts are developed balancing wood and non-wood production, while maintaining economic attractiveness of forests. The information presented may be of interest to scientists as well as to forest managers and may hopefully contribute to sound decision making in forest management.

2. State of the Art

2.1 Future Prospects for the Production of Timber from Valuable Broadleaves

Peter Savill[1], Gary Kerr[2] and Marijan Kotar[3]
[1]*Oxford Forestry Institute, University of Oxford, Oxford, England*
[2]*Forest Research Agency, Alice Holt Lodge, Surrey, England*
[3]*Department of Forestry, University of Ljubljana, Slovenia*

Abstract

In Europe, the trees that could provide the basis of much of the heterogeneity and diversity sought by the public and media are the valuable broadleaves. The valuable broadleaves are comparatively site-demanding and most are light-demanding. Many of the species have the potential to produce high quality timber on relatively short rotations. This chapter examines the prospects for the production of timber from valuable broadleaves in the context of the increasing demands being placed on European forests to produce a broad range of environmental and social benefits. This is achieved by posing two questions: firstly, to what extent does the fulfilment of a wider group of forest management objectives compromise timber production from valuable broadleaves? Secondly, how can valuable broadleaves contribute towards increasing stand heterogeneity and the achievement of the 'new' wider group of forest management objectives in European woodlands?

Keywords: valuable broadleaves, site, silvicultural systems

2.1.1 Introduction

The term 'valuable broadleaves' is often used when discussing the main species in the group, i.e. common ash (*Fraxinus excelsior* L.), wild cherry (*Prunus avium* L.) and sycamore (*Acer pseudoplatanus* L.). However, valuable broadleaves also include species from the genera *Alnus*, *Carpinus*, *Castanea*, *Juglans*, *Malus*, *Pyrus*, *Sorbus*, *Tilia* and *Ulmus* (Turok et al. 1996). In this chapter the main species considered are ash, cherry and sycamore with some specific references to other valuable broadleaves. Valuable broadleaves usually grow in rather a scattered manner, in small clumps or groups, in mixed forests. Typically they make up less than 5% of the forest cover, being competitive only on comparatively rare suitable sites, where they can produce high-quality timber. Elsewhere these species tend only to be found on sites where no other major species will thrive. This may be part of the reason why the valuable broadleaves are an overlooked group of trees compared with oak (*Quercus robur* L. and *Q. petraea* (Matt.) Lieb.) and beech (*Fagus sylvatica* L.).

Recent changes in forestry policy throughout Europe have been driven by the outcomes of the 1992 United Nations Commission on Economic Development (UNCED) summit in Rio, and the Helsinki Ministerial Conference in 1993 (Forestry Commission 1998). The former was mainly concerned with the maintenance and enhancement of biodiversity and the latter with the sustainable management of forests in Europe. However, what both these agreements indicate is that the pressure on forest resources, in Europe and throughout the world, is increasing and is reflected in the increasingly complex sets of objectives for forest management. This requires an innovative response from forest managers. Examples of how the forest industry has responded include a greater interest in mixed species forests (Bartelink and Olsthoorn 1999) and increasingly widespread use of forest management based on natural processes (Pro Silva 1996).

Historically, the most important forest management objective for the majority of forests in Europe has been the production of timber. Against the background of increasing pressure on European forest resources this Chapter aims to examine the prospects for the production of timber from valuable broadleaves. This is achieved by posing two questions: firstly, to what extent does the fulfilment of a wider group of forest management objectives compromise timber production from valuable broadleaves? Secondly, how can valuable broadleaves contribute towards increasing stand heterogeneity and the achievement of the 'new' wider group of forest management objectives in European woodlands?

2.1.2 Objectives of forest management

Historically forests throughout Europe have been managed using silvicultural systems where stand structure has been optimised for timber production (Matthews 1989). However, some recent silvicultural guides have recognised that the silviculture of broadleaved woodland usually has many objectives and have produced guidance on where conflicts between different objectives may be small or large. One such guide has been produced by Evans (1984) from which the information in Table 2.1 is taken. Close investigation of Table 2.1 shows that where the primary objective of the woodland is commercial timber production, interactions with a broad range of other environmental and social objectives are small, or at worst only locally important. For woodlands where commercial timber production is a secondary objective, interactions are again small but possibly significant where recreation and specific nature conservation objectives are of prime importance.

A second model can also be studied to answer the question 'to what extent does the fulfilment of a wider group of forest management objectives compromise timber production from valuable broadleaves'. This is a simple, visual representation of the Helsinki guidelines (1993) on the sustainable management of forests in Europe (Figure 2.1). This demonstrates that the main pillars of sustainable forest management are social, environmental and production factors, with timber production being an important part of the latter. This second model emphasises that timber production is an intimate part of sustainable forest management and not a

Table 2.1. Objectives of growing hardwoods and how they interact (from Evans 1984).

	Designated primary objective							
	Commercial timber production	Minor products/ estate needs	Landscape	Recreation	Sporting	Farm woods/ shelter	Conservation general	Conservation specific
Secondary objective								
Commercial timber Production	–	*	**	***	**	*	**	***
Minor products/estate needs	*	–	*	**	*	*	*	**
Landscape	*	*	–	*	*	**	*	*
Recreation	*	*	*	–	**	**	**	***
Sporting	**	**	*	***	–	*	*	**
Farm woods/shelter	**	*	**	***	*	–	**	***
Conservation:								
General	**	*	*	*	*	*	–	–
Specific	**	**	*	***	**	***	–	–

Legend Effect on main objective of integrating secondary one:
* little effect, easily reconcilable;
** may locally restrict carrying out of main objective;
*** greatly affect main objective, substantial compromise and considerable care needed to achieve both objectives.

Note: Conservation: general – encouragement of habitats favourable to wildlife;
specific – conservation of rare species or woodland ecotype.

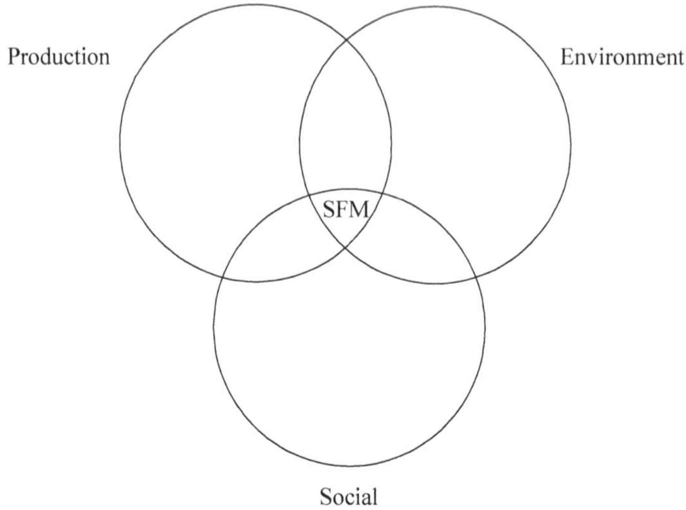

Figure 2.1. A model of sustainable forest management (SFM).

'bolt-on' extra. There is broad agreement between the sustainable forest management model and the Evans' (1984) model, that only small compromises are required to integrate timber production with a broader range of forest management objectives. Hence the answer to our first question is, perhaps rather predictably, that pursuit of a wider range of forest management objectives need only compromise timber production from valuable broadleaves to a very small extent.

Arguably the more interesting question is the second, 'how can valuable broadleaves contribute towards increasing stand heterogeneity and the achievement of the 'new' wider group of forest management objectives in European woodlands?' However, in order to answer this it is important to understand the main silvicultural and ecological characteristics of the valuable broadleaves, their relationship with traditional silvicultural systems and how this may change in the future as forest management becomes increasingly multi-objective.

2.1.3 Characteristics of the valuable broadleaves

Site-demanding nature

One of the more common beliefs about the valuable broadleaves is that they are nutrient-demanding (Miller 1984) in comparison with more widespread species such as oak and beech. They tend only to grow on sites that are perceived as 'fertile'. Walnut (*Juglans regia* L.), ash and cherry, in particular, have this reputation. However, Miller (1984) argues persuasively that they do not require more nutrients to grow to a particular size than any other tree species. Rather, they appear to require rather specific soil and nutritional conditions for optimum growth.

They need their nutrients to be in an easily available form, and tend therefore to be found only on soils that are considered 'fertile'. The species are therefore no more nutrient-demanding than any others, but they are 'site-demanding'. The sites where they grow best are often the ones that have been converted over the centuries into arable or grazing land, so the availability of suitable sites for them has been much reduced by human activity.

As a consequence of this the valuable broadleaves have another common characteristic: a relatively limited capacity for competition in forest stands where sites are not entirely favourable. Where the oaks will produce economically valuable trees over a wide range of climatic conditions and on very varied soils, ranging from acid to alkaline, and from quite dry to wet, the range of sites suitable for the valuable broadleaves tends to be much narrower. A glance at publications such as Pyatt et al. (2001), which list the suitability of sites for different species in terms of a range of climatic and soil factors in the United Kingdom, confirms this. Optimum sites for some of the valuable broadleaves (such as ash, cherry and walnut) are comparatively rare and small in extent. This partly explains the scarcity of the species.

Light-demanding nature

The valuable broadleaves, with the exception of lime (*Tilia* spp.) and hornbeam, are also light demanding in their mature forms and can have difficulty establishing themselves in closed forest conditions, where canopy disturbance is infrequent. However, some species such as ash and sycamore have seedlings which are quite shade tolerant and these can persist in woodlands for many years (Tapper 1992, 1993). Regeneration of the strongest light demanders, such as alder and walnut, and the species that grow rather slowly, or not to very large maximum sizes (those in the Rosaceae), can scarcely survive at all in closed forests. This characteristic largely determines the silvicultural system in which they can be used. Many are found on woodland edges. Among the valuable broadleaves, the maximum heights found in Britain (Mitchell 1978) are ash (45 m), sycamore (35 m), cherry (30 m), alder (22 m), chestnut (35 m), walnut (23 m), apple (10 m), *Sorbus* (~20 m), lime (~32 m), elm (*Ulmus glabra*) (38 m). Of all these, only ash can exceed the maximum height of beech (40 m), and sycamore and sweet chestnut can equal the height of oaks (~35 m).

Life spans of valuable broadleaves

The British Arboricultural Association (1991) has produced a list of life expectancies of trees in six categories. Although some trees can survive for many centuries, most live for no more than 250 years and some are unlikely to survive for more than 50 or 60 years before they start to die back or shed branches. Damage caused by poor pruning, impacts and soil compaction by vehicles, deer, and a variety of other causes can reduce life spans considerably. Under garden or parkland conditions the oaks, limes, sycamore and sweet chestnut (*Castanea sativa* Mill.) will live for 200–300 years; ash and walnut for up to 150 years; those in the Rosaceae for a maximum of 100 years, and alder for only 70 years. Life expectancies in forest conditions might often be somewhat less. Thus, with the exception of the limes, sycamore and sweet chestnut, most of the valuable broadleaves are rather short-lived trees.

Summary

General silvicultural guidelines are difficult to develop for the valuable broadleaves because they are a heterogenous group and species reactions will differ between management regimes and site conditions (Rotach 1999); however, particular characteristics usually noted are given in Table 2.2 (mostly summarised from Kotar and Maucis 2000, Weber and Bahr 2000, Kotar 1998, Franc and Ruchard 1996, Savill 1991, and Evans 1984). One feature common to most valuable broadleaves is that on the majority of sites they require more regular silvicultural interventions, for both survival and adequate development in comparison to many other species. Their promotion and subsequent management needs to take into consideration their particular requirements. Regular release from competition by quite heavy thinning, in order to keep the crowns free, is particularly important. The following section attempts to quantify this fact using some recent research data from studies at the Oxford Forestry Institute.

2.1.4 Thinning valuable broadleaves

The aim of thinning is normally to improve the quality of a stand, but different species need to be thinned according to their particular characteristics and space requirements. An indication of the space needed by the various valuable broadleaves for which data are available can be determined from the very strong linear relationships that are found between the diameter of the crown of a species and the diameter of its stem at breast height (see Table 2.3 and Figure 2.2). Some other species (oak, larch (*Larix* spp.), beech and Norway spruce (*Picea abies* (L.) Karst.) are included in the table for comparison. The practical value of knowing the relationship is that if, by contact with neighbouring crowns, a tree's crown is limited in size, then the diameter of its stem will be correspondingly restricted. In general, the species at the top of Table 2.3 are those whose requirements for space increase most quickly as they grow; they are the commonly acknowledged light-demanders, while those at the bottom are shade bearers.

The equations in Table 2.3 can be used to calculate the expected crown diameters for any stem diameter or, if the stand mean diameter is known, then stand mean crown diameter can be predicted. The number of stems per hectare corresponding to various levels of canopy closure can then be estimated. An example is given in Table 2.4, which shows the number of stems per hectare after thinning to 65% canopy closure, and the equivalent space (in m^2) required per tree, based on the regressions in Table 2.3. In order to obtain good diameter growth, it is suggested that the stocking is never allowed to rise above values corresponding to 100% canopy closure. If a thinning cycle over 6 years is used, then thinnings will have to be very heavy. This may mean reducing the canopy closure to below 65% (Pryor 1988). Most importantly, however, Table 2.4 shows that the valuable broadleaves except for cherry, sycamore and lime need much more space than, for example, beech and oak, and consequently will only grow well in adequately thinned stands. An ash tree of 40 cm dbh needs 53% more space in which to grow than a beech of the same size, and a walnut tree, a 94% greater area. Unfortunately, in practice it is rare that thinning is carried out heavily enough to accommodate scattered valuable hardwood species.

2.1.5 Traditional silvicultural systems

A decision tree for the main high forest silvicultural systems is shown in Figure 2.3 and the following notes describe the suitability of each of the main systems, and coppice-with-standards, to the species characteristics of the valuable broadleaves.

Single tree selection systems
Generally only shade tolerant species are suitable for single tree selection systems, which aim to perpetuate a continuous cover of trees (e.g. of the broadleaved species only beech, hornbeam and lime). If a selection system is introduced in stands that contain light-demanding (intolerant) species the composition will change over time as they become replaced by more shade tolerant species. Eventually, in the absence of any major disturbance, the stand will become dominated by the most tolerant species capable of growing on the site. Of the valuable broadleaves, only lime and hornbeam are really suited to this system. Single tree selection systems are not easy to achieve in practice (Schütz 1999). If the aim is to grow a reasonably high proportion of light-demanding species then, in managing the forest, systems that involve different scales and patterns of canopy disturbance at each intervention are necessary, such as group selection.

Group selection systems
These are the closest approximation to the processes that occur in truly natural forests in temperate Europe, where the main disturbance is caused by wind (Peterken 1996; Jones 1945). They involve the creation of small gaps, typically of 0.1-0.3 ha (circles with diameters of 35 to 66 m). These provide light conditions that are suitable for all but the most intolerant species where groups of at least 0.5 ha (80 m diameter circles) would be needed. Peterken (1996) states that the creation of such large groups is pressing the definition of 'selection' to its limits. Group selection systems have been extensively applied in the past, in central Europe, especially in Austria and Bavaria where groups smaller than 0.1 ha have typically been applied with shade-bearing species, particularly beech, silver fir and Norway spruce.

Advocates of group selection say that it can be sufficiently flexible to accommodate intolerant species such as oak and pine (Everard 1986). The most useful size range is probably 0.1 to 0.5 ha, the larger groups being needed in taller and more uneven aged stands, and for light-demanding species. Among the valuable broadleaves, apart from the limes mentioned above, only sycamore could qualify as a species that could cope with anything but the largest group size. The other species (ash, cherry, alder, chestnut, walnut, and those in the Rosaceae) are probably too light-demanding to be able to thrive well in the establishment and juvenile stages, except in the largest groups of circa 0.5 ha.

Table 2.2. Summary of the silvicultural characteristics of the valuable broadleaves. (Sources: Kotar and Maucis 2000; Weber and Bahr 2000; Kotar 1998; Franc and Ruchard 1996; Savill 1991; Evans 1984).

Species	Drought tolerance[1]	Shade tolerance	Climatic and site preferences	Sites to avoid
Acer pseudoplatanus (sycamore)	Moderate	High as seedling, then intermediate	Climate not usually a limitation. Needs moist, fertile, free-draining soils >pH 4.5.	Avoid sites that are too dry or too wet; pH <4. Podzols, clays gleyed close to surface.
Alnus spp (alder)	Intolerant	Very low	No major climatic limitations, but best in wetter regions. Soil type not a major problem. Best by lakes, streams and on soils with restricted internal drainage with >pH 6. Natural pioneers due to N-fixing ability.	Infertile soils, drought-prone-soils, acid peats.
Castanea sativa (sweet chestnut)	Tolerant	Low	Warm summers needed, so best in southern temperate Europe. Frost- and exposure-tender. Best on deep, fertile, light soils, pH 4-4.5 with ample but not excessive moisture.	Avoid badly drained sites and heavy soils.
Fraxinus excelsior (ash)	Tolerant	High as seedling, then very low	Does best in milder, moister, sheltered regions. Frost tender. Grows best on moist, deep, well drained loams (pH 7-8) with high available N and P and high base saturation. Really suitable sites are rare and small in area.	Avoid acid (pH<4.4) and dry sites, heathlands and heavy, wet soils.

Table 2.2. Continued.

Juglans regia (walnut)	Tolerant	Very low	Warmth and a long growing season needed, so best in southern parts of Europe. Best soils are fresh, rich, loose loams with low water tables and good drainage, pH 6.5-7.5. Very site-demanding. Suitable sites are rare and small in area.	Avoid light sandy soils, heavy soils, shallow soils over chalk, peaty soils, and damp situations generally.
Prunus avium (cherry)	Moderate	Low to very low. Does not tolerate lateral shading	Does best on moist, deep, often clayey soils of near neutral pH, typically deep flushed soils on lower valley slopes.	Will not tolerate exposure. Avoid shallow soils, sands and wet areas.
Sorbus aucuparia (rowan or mountain ash)	Tolerant	Low	Very hardy; no climatic limitations. Does best on light acidic soils.	
Sorbus torminalis (wild service tree)	Tolerant	Low	A rare tree, more common in more southern parts of temperate Europe. Does best on similar sites to cherry, also on dry calcareous soils, but must be nutrient-rich to grow to large sizes.	Similar to *Prunus avium.*
Tilia cordata (small-leaved lime)	Tolerant	High	A tree of the lowlands and lower ranges of hills; production of viable seed limited by cold in northern parts of range. Grows well on a wide range of soils in drier regions, but on limestone in wetter ones.	No major soil limitations, but avoid very dry and very acid soils, and peats.
Tilia platyphyllos (large-leaved lime)	Moderate	High	Does best on rendzinas derived from limestone, or basic igneous rocks, otherwise similar to *T. cordata.*	

1 mostly from Ellenberg (1988)

Table 2.3. Linear regressions of crown diameter (y) against stem diameter (x) for various species ordered according to the slope of the regression. Data were collected in southern England and Wales from dominant canopy trees.

Species	No. of observations	Regression Slope	Intercept	r^2
Castanea sativa	43	0.204	2.793	0.75
Fraxinus excelsior	61	0.204	0.758	0.93
Populus spp	37	0.190	0.801	0.97
Acer pseudoplatanus	58	0.187	0.692	0.89
Sorbus aucuparia	13	0.184	1.096	0.93
Juglans regia	46	0.182	2.750	0.93
Larix kaempferi	31	0.174	0.847	0.97
Quercus petraea & *Q. robur*	63	0.168	1.134	0.94
Prunus avium[1]	48	0.158	1.190	0.88
Fagus sylvatica	58	0.152	1.133	0.92
Tilia cordata	33	0.114	0.692	0.90
Picea abies	50	0.106	0.141	0.88

1 From Pryor (1988)

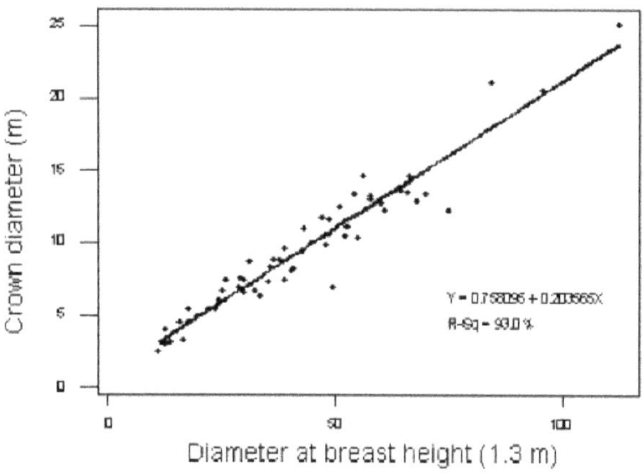

Figure 2.2. Relationship between crown diameter and stem diameter of ash (*Fraxinus excelsior* L.).

Shelterwood systems

These systems are particularly appropriate, on suitable sites, for regenerating various moderately to strongly light-demanding species, such as most of the valuable broadleaves. These often fail to regenerate on clear cut sites unless advance regeneration is already present (Marquis 1982). The failure to establish sufficient regeneration in the open can be caused by inadequate seed dispersal,

Table 2.4. Number of stems per hectare after thinning to 65% canopy closure (CC), and the equivalent space (in m^2) required per tree, based on the regressions in Table 2.3. The trees are ordered in the table according to their space requirements.

	20 cm dbh		30 cm dbh		40 cm dbh		50 cm dbh		60 cm dbh	
	Stems/ha at 65% CC	Space per tree (m^2)	Stems/ha at 65% CC	Space per tree (m^2)	Stems/ha at 65% CC	Space per tree (m^2)	Stems/ha at 65% CC	Space per tree (m^2)	Stems/ha at 65% CC	Space per tree (m^2)
Castanea sativa	176	57	104	96	69	145	49	205	37	274
Juglans regia	202	49	123	82	82	122	59	169	45	224
Fraxinus excelsior	353	28	175	57	104	96	69	145	49	205
Sorbus aucuparia	363	28	189	53	115	87	-	-	-	-
Populus spp	391	26	196	51	117	85	78	128	55	181
Acer pseudoplatanus	422	24	208	48	124	80	82	122	59	171
Quercus petraea & robur	410	24	217	46	134	75	91	110	66	152
Larix kaempferi	442	23	225	44	136	74	91	110	65	154
Prunus avium[1]	437	23	236	42	147	68	100	100	72	138
Fagus sylvatica	475	21	255	39	159	63	109	92	79	127
Tilia cordata	937	11	489	20	300	33	202	49	146	68
Picea abies	1619	6	751	13	431	23	280	36	-	-

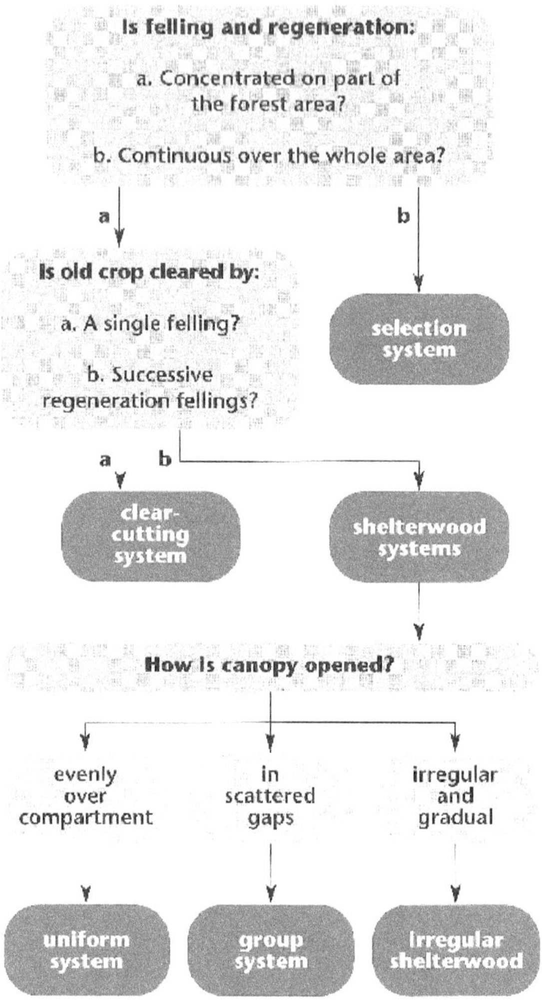

Figure 2.3. A decision tree for classifying the main high forest silvicultural systems, adapted from Kerr (1999). Note that selection systems can be further subdivided into group and single-tree selection depending on the shade tolerance of the species involved.

excessive competition with weeds, frost damage or excessive deer browsing. Appropriate sites for shelterwoods are generally well drained and have a good overstorey stocking of desirable species of seed bearing age. Unsuitable sites are poorly drained, or have a low overstorey stocking, or an overstorey with predominantly undesirable species. In the seeding felling, if too high a basal area is removed, some seedlings will grow very fast but too few tend to become established. If too little is removed, seed germination and establishment is reduced and mortality becomes excessive as the canopy closes, so that few seedlings survive more than 3–5 years, except for the most shade tolerant species.

Coppice and coppice-with-standards

This system has declined since the mid-1800s from being, probably, the most commonly used of all silvicultural systems in broadleaved forests to a state of virtual obsolescence. This has happened during the lifetime of broadleaved trees that are now mature and still standing in forests all over Europe. The decline has been caused by industrialisation (manufactured alternatives to products formerly made from coppice materials, such as barbed wire), and the availability of alternative sources of fuel (coal, oil, gas and electricity). Both have had the effect of reducing the need for coppice products to very low levels, and so the system has become economically unviable. Coppicing is particularly good at promoting many of the valuable broadleaves, including lime.

The decline of the coppice-with-standards system in Europe has resulted in large abandoned areas being present in most broadleaved regions. It is now difficult to see the succession of ages of oak standards, and where any still exist, they all appear old and large (Peterken 1981). Where the interval between annual cuts lengthened, rotations inevitably lengthened. Unwanted species were no longer removed as coppicing declined and the freeing of any young seedlings or saplings was neglected. The failure to release regularly light-demanding trees from competition has probably contributed significantly to the comparative rarity of *Sorbus torminalis* and most other valuable hardwood species, which in the more shaded conditions are no longer able to compete successfully with their more shade-bearing neighbours.

Thus, in general, apart from the limes, the most appropriate silvicultural systems for the valuable broadleaves are group selection, using very large groups of 0.5 ha or, on appropriate sites, shelterwood systems, or coppice-with-standards.

2.1.6 Future silvicultural systems

The traditional silvicultural systems described in the last section have three major limitations for allowing the demands for forest management to meet an increasingly wide range of objectives (Coates and Burton 1997). Firstly, silvicultural systems are often thought of as a prescibed set of treatments which are applied without thought and consequently application of procedures is often very rigid. Secondly, traditional descriptions of silvicultural systems focus on the trees and their relationship with timber production. Commonly systems are described in terms of the method of reproduction, basal area, maximum tree size etc. This provides an incomplete framework for greater understanding of the woodland as an ecosystem, which will be increasingly important for achieving many social and environmental forest management objectives. Thirdly, most silvicultural systems encourage stand homogeneity in order to optimise and predict tree growth. However, all forest stands have a degree of heterogeneity that may need to be exploited to achieve a wider range of forest management objectives.

Hence, continued use of the traditional silvicultural systems of clear-cutting, shelterwood and selection, may be convenient but may also preclude innovative thinking (Aplet et al. 1994). Thinking in North America (O'Hara et al. 1994) is encouraging the view that silvicultural systems should be viewed as a structural gradient from clear-cut to single tree selection, with each point along the gradient representing an opportunity to vary the elements to meet different combinations of structural objectives. This type of thinking was first encouraged by Smith (1972) who pleaded for 'elegant variation' in the evolution of silvicultural treatment and for it to be 'more varied than our wildest present imagination can encompass'. Changes of this kind will be required by forest managers throughout Europe to meet the increasingly complex demands being made on the resources for which they are guardians.

Under this future scenario, elements of stand heterogeneity, often represented by the presence of the valuable broadleaves should be encouraged. This is in contrast to the existing situation where only the application of certain rigidly defined combinations of silvicultural systems and sites favour the valuable broadleaves. The future situation may be that wherever valuable broadleaves exist, or have the potential to regenerate, silviculture should aim to encourage them, predominantly by a more flexible approach to the management of gaps (Coates and Burton 1997). The guidance in Tables 2.2 and 2.3 of this Chapter will help define management strategies to encourage valuable broadleaves within European woodlands.

One of the main challenges to this vision of the future is that, as Schütz (1999) has pointed out, mixtures of tree species complicate management and stand development. They need more frequent, and therefore expensive, silvicultural interventions than pure stands to ensure the survival of the less competitive species. If these interventions are not undertaken in a timely manner, the forest usually becomes dominated by very few species. Generally, one of the biggest impediments to implementing the sort of silviculture that would promote valuable broadleaves, and ecologically sensitive management is the cost. There is an increasing divergence between economic and ecological interests in forestry, and one of the most pressing needs today is to find systems that can bring the two together.

In summary, how can valuable broadleaves contribute towards increasing stand heterogeneity and the achievement of the 'new' wider group of forest management objectives in European woodlands? The answer appears to be that if innovative ways are to be found to manage for multiple objectives then this will probably result in harvesting at different scales within the same species/site unit. This should allow valuable broadleaves a greater chance to survive and grow in European woodlands and hence make a significant contribution to timber production, diversity and the social values.

References

Aplet, G.H., Sampson, R.N. and Adams, D.L. 1994. Beyond even vs. uneven-aged management: toward a cohort-based silviculture. Journal of Sustainable Forestry 7(2): 38–47.

Arboricultural Association 1991. Arboricultural Handout No. 4, Tree Management. Arboricultural Association, Ampfield, Romsey, Hampshire, England. 2 p.

Bartelink, H.H. and Olsthoorn, A.F.M. 1999. Introduction: mixed forest in western Europe. In: Olsthorn, A.F.M. , Bartelink, H.H., Gardiner, J.J., Pretzsch, H., Hekhuis, H.J., and Franc, A. (eds.). Management of mixed-species forest: silviculture and economics. IBN Scientific Contributions 15. IBN-DLO, Wageningen, Holland. Pp. 9–16

Coates, K.D. and Burton, P.J. 1997. A gap-based approach for development of silvicultural systems to address ecosystem management objectives. Forest Ecology and Management 99: 337–354.

Ellenberg, H. 1988. Vegetative ecology of central Europe. Cambridge University Press, Cambridge, UK.

Evans, J. 1984. Silviculture of broadleaved woodland. Forestry Commission Bulletin 62.

Everard, J.E. 1986. Management of broadleaved forests in western Europe. Forestry Commission, Edinburgh. Unpublished. 139 p.

Forestry Commission 1998. The UK Forestry Standard. Forestry Commission, Edinburgh.

Franc, A. and Ruchaud, F. 1996. Autecologie des feuillus precieux: frene commun, merisier, erable sycomore, erable plane. [Autecology of high-value broadleaved trees: ash, wild cherry, sycamore maple and Norway maple.] Etudes-du-CEMAGREF:-Serie-Gestion-des-Territoires, No. 18. Centre National du Machinisme Agricole, du Genie Rural, des Eaux et des Forets (CEMAGREF); Riom; France. 170 p.

Jones, E.W. 1945. The structure and function of the virgin forest of the north temperate zone. New Phytologist 44: 130–148.

Kerr, G. 1999. The use of silvicultural systems to enhance the biological diversity of plantation forests in Britain. Forestry 72(3):191-205.

Kotar, M. 1998. Distribution and growth characteristics of the wild service tree in Slovenia. Gozdarski Vestnik 56: 258–278. In Slovenian.

Kotar, M. and Maucis, M. 2000. Wild cherry – an important tree species in the Slovenian forests. Gozdarski Vestnik 58: 227251. In Slovenian.

Marquis, D.A. 1982. Effect of advance seedling size and vigour on survival after clear-cutting. USDA Forest Service Research Paper NE-498.

Matthews, J.D. 1989. Silvicultural systems. Oxford University Press.

Miller, H.G. 1984. Nutrition of hardwoods. In: Report of fifth meeting of National Hardwoods Programme. Commonwealth Forestry Institute, Oxford. Pp. 17–29.

Mitchell, A.F. 1978. A field guide to the trees of Britain and northern Europe. Collins. 416 p.

O'Hara, K.L., Seymour, R.S., Tesch, S.D. and Guildin, J.M. 1994. Silviculture and our changing profession: leadership for shifting paradigms. Journal of Forestry 92(1): 8–13.

Peterken, G.F. 1981. Woodland conservation and management. Chapman and Hall, London.

Peterken, G.F. 1996. Natural woodland ecology and conservation in northern temperate regions. Cambridge University Press, Cambridge, UK.

Pro Silva 1996. Pro Silva. Association of European foresters practising management which follows natural processes, Strasbourg, France.

Pryor, S.N. 1988. The silviculture and yield of wild cherry. Forestry Commission Bulletin 75. HMSO, London.

Pyatt, G., Ray, D. and Fletcher, J. 2001. An ecological site classification for forestry in Great Britain. Forestry Commission Bulletin 124. HMSO, London.

Rotach, P. 1999. In situ conservation and promotion of noble hardwoods in Europe through improved silvicultural management strategies. Forest – Genetic Resources 27: 43–48.

Savill, P.S. 1991. The silviculture of trees used in British forestry. CAB International, Wallingford, England.

Schütz, J.-P. 1999. Close-to-nature silviculture: is this concept compatible with species diversity? Forestry 72: 359–366.

Smith, D.M. 1972. The continuing evolution of silvicultural practice. Journal of Forestry, 70(2): 89–92.

Tapper, P.G. 1992. Demography of persistent juveniles in *Fraxinus excelsior*. Ecography 15: 385–392.

Tapper, P.G. 1993. The replacement of *Alnus glutinosa* by *Fraxinus excelsior* during succession related to regenerative differences. Ecography 16: 212–218.

Turok, J., Eriksson, G., Kleinschmit, J. and Cangar, S., compilers 1996 Noble hardwoods network: report of first meeting 24–27 March 1996, Escherode, Germany. International Plant Genetic Resources Institute (IPGRI), Rome Italy.

Weber, G. and Bahr, B. 2000. Eignung bayerischer Standorte fur den Anbau von Esche (*Fraxinus excelsior* L.) und Bergahorn (*Acer pseudoplatanus* L.). [Suitability of Bavarian forest sites for the cultivation of ash (*Fraxinus excelsior* L.) and sycamore (*Acer pseudoplatanus* L.).] Forstwissenschaftliches-Centralblatt 119(5): 263–275.

2.2 Results of a Questionnaire on Management of Valuable Broadleaved Forests in Europe

Michael Thies, Sebastian Hein and Heinrich Spiecker
Institute for Forest Growth, University of Freiburg, Germany

Abstract

The results of a written questioning of experts on the importance and management of valuable broadleaved forests on a European level are presented. The analysis focuses on common ash, sycamore and wild cherry which are considered to be the most important representatives of the group of valuable broadleaved tree species. In respect to economical importance and proportion of the area, the questioned experts expect a significantly increasing role of the above mentioned tree species. In addition, whereas currently the first 10 m of stem wood are mainly used for saw wood, future production will strictly be focused on high value veneer logs. The analysis of the description of production targets, based on the variables production time, target diameter, length of the branch-free bole and number of trees per hectare in the final stand, demonstrates that only for wild cherry adapted target combinations are common. Although for ash and sycamore a wide range of target combinations can be observed, only minor differences between ash and sycamore are negligible. Stands of ash, sycamore and wild cherry are mainly regenerated naturally. Positive selection is the most important treatment in young stands. The majority of experts prefer future crop tree thinning, where across most of Europe growth control and thinning regimes are determined according to stand level criteria such as the number of trees in the stand. In contrast, German future crop tree thinning is now focusing on the retaining of the crown base at a certain height or on guaranteeing crown development without any competition. Pruning of wild cherry is common whereas pruning of ash and sycamore, particularly in Germany, is not part of the regular stand treatment. In conclusion, it appears that there is a need for information on forest management systems which assure the production of high valuable timber.

Keywords: questionnaire, valuable broadleaved forests, production targets, Acer pseudoplatanus L., Fraxinus excelsior L., Prunus avium L., tending, thinning

2.2.1 Introduction

Valuable broadleaved species are of increasing interest for forest management on a European level for many reasons (cf. Appendix 1 in this book). Although methods of managing valuable broadleaved forests are known (e.g. Joyce et al. 1998; Kerr and Evans 1993; Spiecker 1994; Thill 1975) on a European level it is insecure whether these concepts meet the information needs of forest practitioners and their future management targets. Thus, to document the recent status quo and derive a systematical overview of common management strategies for stands dominated by valuable broadleaved tree species in the time period from 1998 to 2000, experts of this field were questioned. The main objective of this survey was to summarize information about the following fields of interest:

(1) Recent and future shares of valuable broadleaved tree species;
(2) Characteristics and differences in production targets and methods in stands composed of valuable broadleaved tree species as well as conceivable changes in future;
(3) Contacts to experts with knowledge about management strategies and specifically treated stands of valuable broadleaved tree species;
(4) Identifying research needs and information gaps.

The questioning was meant to be the initial work package of a research study funded by the European Forest Institute on "Management of Mixed Valuable Broadleaved Forests in Europe" (ID 92/8553 EFI). Upon analyzing which tree species should be assigned to the group of valuable broadleaved tree species (cf. Appendix 2 in this book), it was found that common ash, (*Fraxinus excelsior* L.), sycamore (*Acer pseudoplatanus* L.) and wild cherry (*Prunus avium* L.) tend to be the major representatives of this group of tree species. For this reason the presentation of results focuses on these tree species.

2.2.2 Design and structure of the questionnaire

The basic methodology of the survey is characterized according to Roth (1995) as a written questioning of a pre-defined group of experts. The selected group of experts consists of heads of forest management districts and enterprises, scientists from universities and research stations as well as representatives of forest ownership associations. The questions were both of an objective and subjective nature. Whereas objective variables (such as forest ownership category, affiliation with a certain group of experts) were mainly used to classify results in the analysis, subjective information questioned specific knowledge, personal opinions, or estimations in respect to management of valuable broadleaved tree species. A second method used to characterize the questions was to differentiate between closed and open questions. We mainly included closed questions (i.e. pre-defined ranks or categories such as often – seldom or small – high) because of fixed decision alternatives and a correspondingly higher efficiency in the analysis of the data. In some cases concrete values to distances, area, size, etc. had to be added. As a result most of the variables are ordinal-scaled. Because anonymity was ensured, all questions could be posed directly.

The questionnaire, which was developed iteratively where each step was tested in respect to explicitness and comprehensibility, can be subdivided into four parts

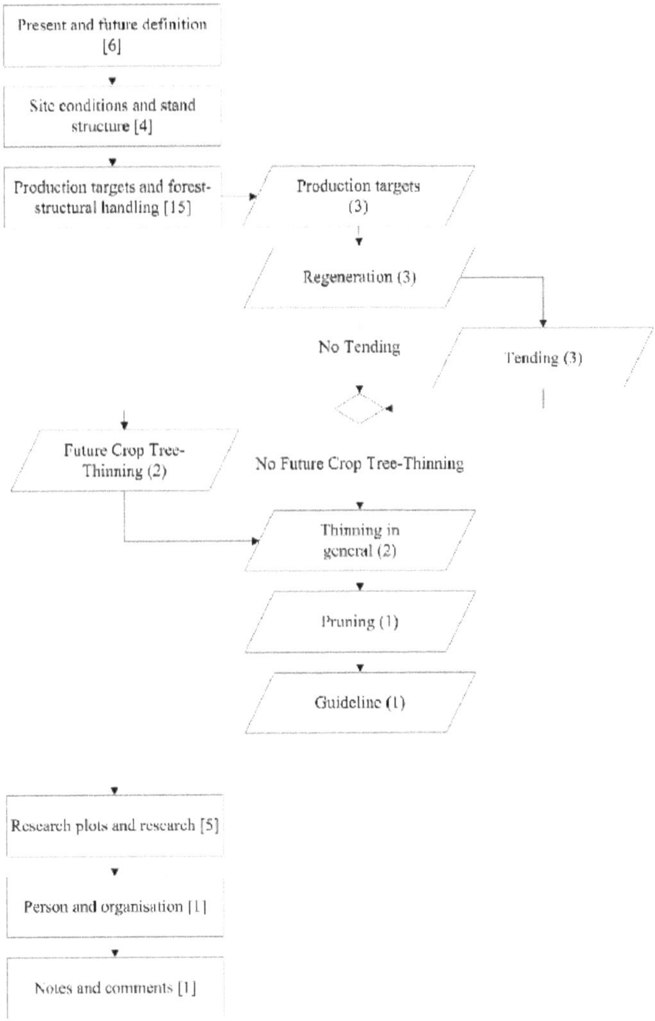

Figure 2.4. Structure of the questionnaire (number of questions shown in brackets).

(Figure 2.4). 1) The main part, about importance, site demands, targets and management consists of 25 questions (see the number of questions per subdivision in brackets). 2) Another five questions address research and established research plots. 3) One sheet is intended for information about the person and the expert's organization and finally 4) the last part gives the opportunity to contribute personal suggestions and comments. The structure of the questions concerning treatment of valuable broadleaved stands includes two conditional alternatives to separate specific methods of treatment that are not of interest to all experts. In the case where tending is not part of the regular treatment, three questions are disregarded. In the next step a separation between future crop-tree thinning and other thinning methods is integrated. As far as no future crop tree thinning is applied, two

Table 2.5. Absolute number of analyzable questionnaires per country.

Country	Analyzable questionnaires	Country	Analyzable questionnaires
Austria	4	Italy	1
Belgium	2	The Netherlands	3
Croatia	2	Poland	3
Czech Republic	3	Romania	2
Denmark	2	Russia	2
Estonia	1	Serbia	1
France	4	Slovakia	1
Georgia	1	Slovenia	1
Germany	46	Spain	4
Greece	1	Switzerland/ Liechtenstein	4
Hungary	2	United Kingdom	3
Ireland	2	Ukraine	4

additional questions can be omitted. Questions on common pruning activities and documented guidelines or decrees conclude the main part of the questionnaire.

Instructions for answering this questionnaire were sent to the individual experts along with background information about the initiated project. Data processing was carried out with the spreadsheet program Microsoft Excel.

2.2.3 Feedback and material for analysis

The questionnaire was separated into two parts. The first focused on Germany, Austria and the German-speaking part of Switzerland. The quota of responses was 53%, i.e. a total number of 51 responded with quality. On the basis of these results the general suitability of the questionnaire was confirmed and only minor revisions were necessary to increase acceptability while guaranteeing comparability with the second part of the questioning which covered a number of 25 other European countries. The quota of returns was 33%, equivalent to 47 questionnaires. Table 2.5 gives an overview of the distribution of the recognized questionnaires in each European country. Due to the limited number of responses per country, an analysis of discrepancies between different European countries is not possible. Moreover, it can be seen that the number of samples from German speaking countries compared to other European countries is significantly higher. To insure that results on a European level are not mainly determined by the results from German speaking countries, the analyses were conducted separately and results were compared. A first overview of the results in the German speaking countries is presented by Thies and Hein (2000). Because the number of analyzable questionnaires differs for each question, in the following the sample size is shown for each figure or important descriptive numerical result.

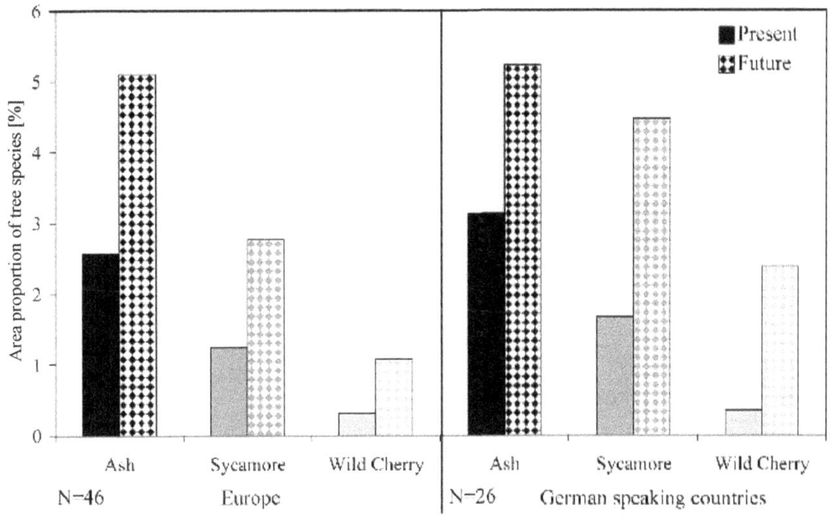

Figure 2.5. Present and projected area proportions of selected valuable broadleaved tree species for German speaking countries and for Europe.

2.2.4 Quantitative results

Recent and future importance of valuable broadleaved species in Europe

In addition to the shares of the valuable broadleaved tree species, the question of the ash, sycamore and wild cherry proportion of the annual total removals in the last three years was asked to characterize the current importance. It turns out that in Europe, excluding Germany, 2.7% ash, 1.7% sycamore and 0.7% wild cherry contribute to the annual total removals. A total of 30 experts answered this question where 53.1% of them referred to statistics and the others estimated the proportion. In German speaking countries the proportions of the total annual removals are lower: 1.4% ash, 0.7% sycamore and 0.4% wild cherry (based on the feedback of 18 experts from which 61.9% referred to statistics and 38.1% estimated the numbers). Focusing on the tree species ash, sycamore and wild cherry, the current and projected proportions are increasing tremendously (Figure 2.5). Compared to the current situation, the area proportion of ash, sycamore and wild cherry in Europe as well as in Germany will double in the long term. For wild cherry in German speaking countries the projected proportion will be three times that of today. Together, the three selected tree species in Europe currently cover 4.1% of the forests (5.1% in German speaking countries). In the future this proportion is expected to reach 8.9% for Europe and 12.1% for German speaking countries. The current economic importance of the tree species for the forest and timber industries is estimated by experts from all European states to be low to moderate. A slight increase, however, is expected within the next two decades. Noticeable differences between the two groups questioned with respect to this subjective estimation of the economic development could not be found. As a main explanation for this increasing importance of valuable broadleaved tree species, the majority of experts point out the high prices for valuable timber produced in short rotations. Additionally, the trend to support mixed stands of deciduous tree species because of their stability, contribution to biodiversity and recreational use, directly contributes to an increasing importance of valuable broadleaved tree species. In some European countries this political goal is accompanied by subsidies (Weber 2000) which support a higher number of these tree species in younger stands and plantings.

Thus, it can be expected in future that younger stands dominated by valuable broadleaved tree species will grow to merchantable dimensions with a corresponding economic value can be observed in several European countries.

Site conditions and stand structure

To characterize the current distribution of ash, sycamore and wild cherry on specific site conditions, nutrient and water supplies were evaluated based on the categories below average, average or above average. Significant differences between Europe (49 responses) and German speaking countries (34 responses) could not be found. The tree species are mainly concentrated on sites with above average nutrient availability. Among all three species, ash (average value 2.49 in Europe and 2.82 in German speaking countries) has higher nutrient demands than sycamore (2.34 and 2.65) and wild cherry (2.4 and 2.5). Wild cherry sites show average water availability. In German speaking countries sycamore stands are found on more moist sites than in Europe (average 2.5 compared to 2.23) whereas for ash, sites with above average water supply are preferred (2.57 in Europe and 2.85 in Germany). In general the results show that all three species are currently growing on sites with high nutrient supplies. In contrast to ash and sycamore, cherry stands do not require a large water supply.

The distribution of age-classes for sycamore and wild cherry is dominated by stands younger than 40 years (Figure 2.6) both in Europe and German speaking countries. Stands of wild cherry that are older than 80 years cannot be observed in any country. For ash, the distribution in German speaking countries is more balanced. In addition to forest districts dominated by younger age classes, those with a high proportion of stands older than 81 years do exist. Outstanding contrasts are seen between the high proportion of cherry stands in German speaking countries which are younger than 20 years and the high number of ash stands in the age class of 41 to 60 years throughout the rest of Europe (Figure 2.6).

53% of the results pertaining to age classes for the European part of the questioning are based on statistics and forest inventory and correspondingly 47% are based on subjective estimations. Of the German speaking group, 63% are based on forest inventory and 37% on personal estimations. Because half of the answers were based on verifiable facts, the conclusion that ash, sycamore and cherry stands in Europe are dominated by young and management intensive age-classes, can be generalized.

Ash, sycamore and wild cherry often grow in mixed stands. The most important tree species mixed with these valuable broadleaves differs between stands of the respective valuable broadleaved species of interest as well as between Europe and

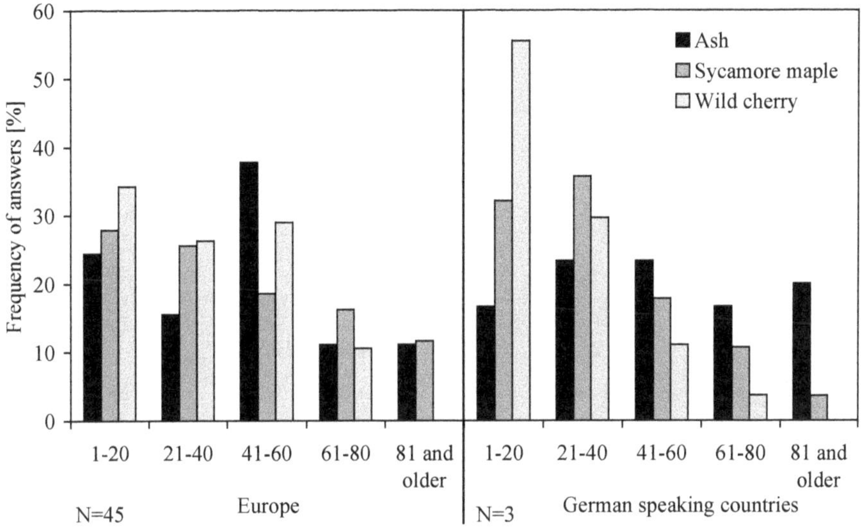

Figure 2.6. Age-class distribution of stands comprised of ash, sycamore and wild cherry.

German speaking countries (see frequency of answers per stand of considered tree species in Table 2.6). For ash stands at a European level, the other tree species most likely to be found is oak, followed by sycamore, alder and beech. In German speaking countries beech is the dominant mixed tree species followed by sycamore, alder and oak. For sycamore stands only minor differences between the two parts of the questionnaire can be observed. Whereas in Europe wild cherry is often the dominant species in a mix, in German speaking countries a mixture with beech, ash lime tree and hornbeam is common.

In most cases the tree species are individually mixed or in small groups of a diameter less than 15 m. Wild cherry in particular is very often individually mixed (82.5% of all responses on the European level and 63.3% for German speaking countries).

For ash and sycamore some of the experts in Europe pointed out that small stands with a diameter of more than 60 m are the predominate form of mixture. Groups (with diameters between 16 and to 60 m) are of minor importance and the frequency of responses for these groupings ranges from 4.5% to 15.6% considering all three tree species.

Common production targets

To obtain an overview of the current utilization of ash, sycamore and wild cherry logs, the experts were asked about the main use of the lower ten meters of the stem and a prognosis on future utilization development of this part of the basal log length. The

Table 2.6. Frequency of the five most important tree species found mixed in stands of ash, sycamore and wild cherry.

Europe	Ash stands	Oak 15.7%	Sycamore 14.0%	Alder 14.0%	Beech 13.5%	Norway maple 5.6%	Others 37.1%
	Sycamore stands	Beech 23.0%	Ash 19.1%	Oak 10.5%	Wild cherry 7.9%	Lime tree 7.2%	Others 32.3%
	Wild cherry stands	Oak 21.1%	Beech 18.0%	Ash 12.0%	Hornbeam 11.3%	Lime tree 8.3%	Others 29.3%
German speaking countries	Ash stands	Beech 25.7%	Sycamore 22.8%	Alder 16.8%	Oak 10.9%	Hornbeam 6.9%	Others 16.8%
	Sycamore stands	Beech 27.0%	Ash 21.0%	Hornbeam 5.0%	Oak 9.0%	Lime tree 8.0%	Others 30.0%
	Wild cherry stands	Beech 22.6%	Sycamore 16.7%	Lime tree 14.3%	Hornbeam 11.9%	Ash 10.7%	Others 23.8%

results show differences between the tree species as well as differences between the two groups questioned (Table 2.7). Whereas in Europe an important use of the first 10 m of the log for all three tree species is veneer production (approx. one third of the experts confirmed this for ash and wild cherry and one fifth for sycamore), in German speaking countries veneer logs are only currently important for wild cherry. In contrast only 12% of the German speaking experts report on the use of sycamore logs for veneer production. Whereas ash veneer production is quite common in Europe, the utilization of ash trees in German speaking countries can be ignored (3% of responses). Currently, the predominant use of these three species across Europe is saw wood. Some experts indicate that industrial and fuel wood is a common use, but this is an exception. Other uses include products that are part of the saw wood group such as parquet flooring, furniture or country specific uses such as ash for hockey sticks in Ireland. The envisaged future use according to the opinion of the experts will change dramatically. While at a European level the proportion of experts preferring logs for veneer production is increasing by one third, in German speaking countries an increase of 60% is expected. Both are mainly connected to a decrease in saw wood production. The use of ash, sycamore and wild cherry by industry and for fuel wood in the future will be of no importance. Because the majority of experts are striving for future veneer log production, silvicultural treatments consequently have to be adapted to focus on quality aspects and quality development of single trees. Otherwise this tremendous rate of increase for the highest paid assortments of veneer logs will not be realized.

Production targets for certain tree species are related to the final stand stage and can be defined by a combination of the target DBH, production time, length of the branch-free bole and the number of trees in the final stand. Stringent production targets should take growth characteristics of the tree species into account as well as interdependencies between the single target components. For instance, an extension of

Table 2.7. Frequency (%) of the current and future primary use of the lower ten meters of stem wood.

[%]			Veneer	Saw wood	Industrial wood	Fuel wood	Other use
Europe	Ash	Present	28.8	63.5	1.9	1.9	3.8
		Future change	+26.9	-21.2	-1.9	-1.9	-1.9
	Sycamore	Present	22.0	70.0	6.0	0.0	2.0
		Future change	+24.0	-20.0	-4.0	0.0	0.0
	Wild cherry	Present	32.6	52.2	2.2	10.9	2.2
		Future change	+28.3	-15.2	-2.2	-10.9	0.0
German speaking countries	Ash	Present	3.0	93.9	3.0	0.0	0.0
		Future change	+60.6	-57.6	-3.0	0.0	0.0
	Sycamore	Present	12.1	84.8	3.0	0.0	0.0
		Future change	+63.6	-60.6	-3.0	0.0	0.0
	Wild cherry	Present	29.0	71.0	0.0	0.0	0.0
		Future change	+58.8	-58.8	0.0	0.0	0.0

production time is consistent with a higher DBH. Simultaneously, a tree with a shorter length of the branch-free bole will have a higher DBH. A high number of trees per hectare is an expression of a high competition between the trees and as a result DBH will be reduced whereas the length of the branch-free bole will increase. With this background the experts were asked to specify production targets for final stands of ash, sycamore and wild cherry. The means of the production values are shown by the grey shaded areas in Figure 2.7. Each axis represents one target parameter. For ash the mean of all target combinations at the European level is characterized by a production time of 84 years (for German speaking countries 106 years), a target diameter of 51.4 cm (58.6 cm), a length of the branch-free bole of 10.1 m (9.2 m) and a density of 157 (109) trees per hectare in the final stand. The optical impression from Figure 2.7 indicates that there are only minor differences between ash and sycamore with respect to mean production targets. This is confirmed by the numerical results: production time for sycamore on average is longer (94 years for Europe, 112 years for German speaking countries) and consequently target diameter is higher (Europe 52.3 cm, German speaking countries 60.3 cm), length of the branch-free bole is marginally lower (Europe 9.5 m, German speaking countries 9.1 m) and finally the number of trees in the final stand of sycamore is lower on a European level (137 trees per hectare) and slightly higher in German speaking countries (111 trees per hectare) than for ash. Significant differences however can be observed for wild cherry. The mean production target in Europe is characterized by a production time of 65 years (78 years in German speaking countries), a target diameter of 46.5 cm (51.8 cm), a length of the branch-free bole of 8.4 m (8.0 m) and a mean density of 124 (94) crop trees per hectare. For wild cherry specific target combinations have been developed and are well known whereas ash and sycamore feature no specifically differentiated production targets.

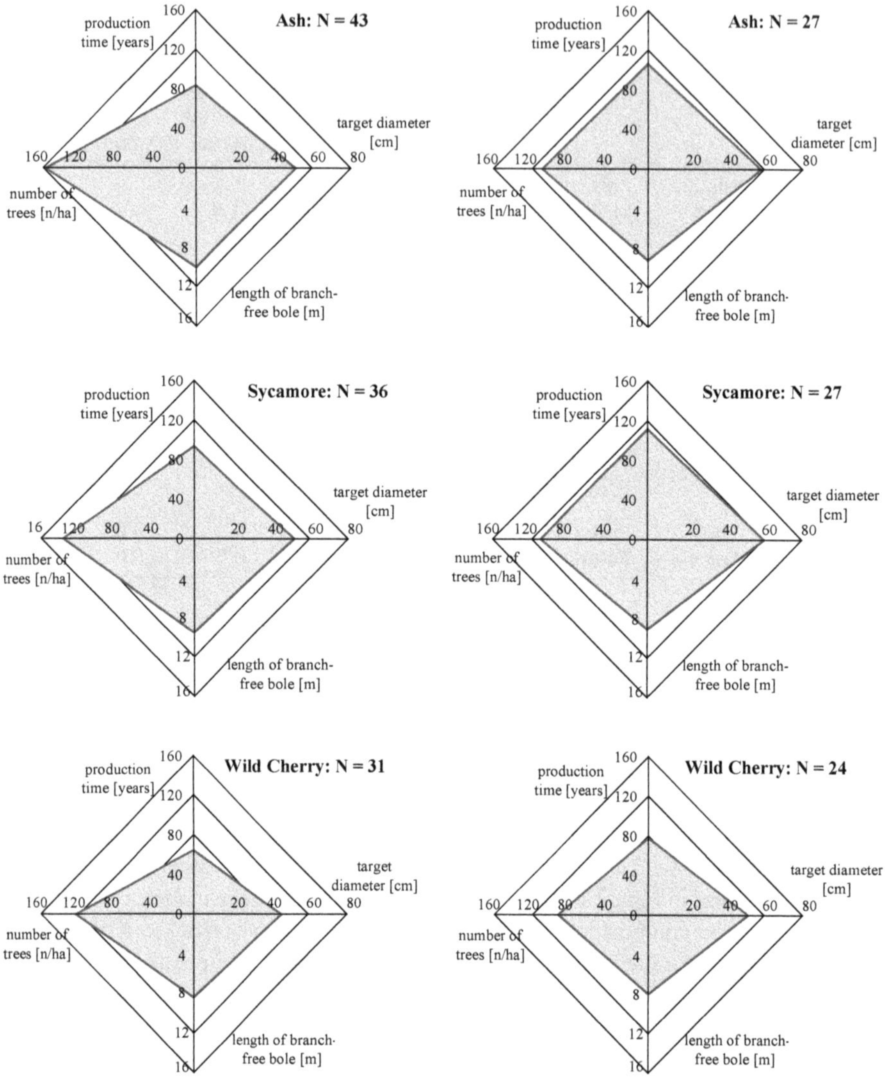

Figure 2.7. Mean of production targets for ash, sycamore and wild cherry in Europe (left) and German speaking countries (right).

Aside from the means of values for the single target parameters, the extreme values are also of interest (Table 2.8). A wide range of parameter combinations can be observed. Again, it can be seen that there are only minor differences between ash and sycamore while clearly separated ranges for the target combinations of wild cherry exist. For example, production time and the number of trees in the final stand are lower for wild cherry than for ash and sycamore, whereas target diameter and the length of the branch-free bole are within the range of the other two tree species.

Table 2.8. Extreme values of production targets for ash, sycamore and wild cherry.

		Production time [years]		Target diameter [cm]		Length of branch-free bole [m]		Number of trees [n/ha]	
		min	max	min	max	min	max	min	max
Europe	Ash	40	200	28	80	5.0	25.0	40	450
	Sycamore	40	350	30	80	4.0	22.0	30	400
	Wild cherry	40	90	18	65	4.0	15.0	20	500
German speaking countries	Ash	70	140	40	70	6.0	18.0	50	300
	Sycamore	70	50	50	90	6.0	18.0	50	200
	Wild cherry	40	120	35	65	5.0	15.0	12	200

Some of the parameter combinations within this range obviously include inconsistencies which are not reflected in the values of the descriptive statistics.

Stand treatment and management

Regeneration

Stands of ash and sycamore in most cases regenerate naturally. According to the experts for ash, 40.6% of the stands within their area of responsibility at the European level are planted (30.9% in the German speaking countries). The proportions for sycamore with 32.6% planted (31.7%) are comparable. In contrast, it is interesting that the regeneration of wild cherry in German speaking countries is mainly planted (74.2%) whereas the proportion at the European level is just 43.5% where natural regeneration dominates with a total amount of 53.8%. Seeding plays no important role for the regeneration of the three tree species; the maximum proportion is 2.7% mentioned by one expert in Europe. For most plantations rectangular planting patterns are recommended. The density of ash and sycamore on average is 2000 plants per hectare at the European level (3000 in German speaking countries) with a range of 200 as the minimum to 25 000 as the maximum (500 to 8000 in German speaking countries); differences between ash and sycamore are negligible. The average density for cherry plantations, however, is lower than for the two other species (1300 and 1800 per hectare in Europe and German speaking countries respectively). 73% of all experts in Europe and 94% of the German speaking experts of valuable broadleaved stands recommend a mixture with a serving tree species. At the European level lime tree and hornbeam are preferred followed by beech. In German speaking countries hornbeam as a serving tree species is of higher importance than lime tree and beech.

Tending of young stands

Most of the experts (85.1% for Europe and 93.3% for German speaking countries) recommend tending young ash, sycamore and cherry stands. Tending young stands

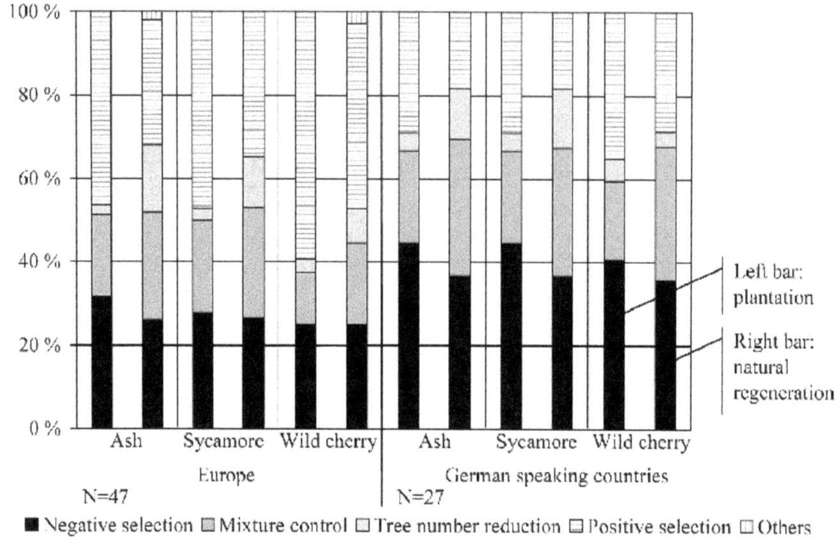

Figure 2.8. Tending activities in young stands depending on the method of regeneration.

often includes negative selection (Figure 2.8). However, most of the authors recommend a combination of negative selection in early stages combined with a mixture control followed by a positive selection in later stages of young stand development. In the positive selection only the few high quality trees that will remain in the stand till the end of the production time are marked permanently. The negative selection system marks all (numerous) competitors to be removed at each thinning. This positive selection in Europe is the dominating treatment of young stands and is more common than in German speaking countries where negative selection is the most important treatment in young stands. Pure systematical tree number reductions play only a minor role and are not often recommended. Comparing the two relevant methods of regeneration (left bar in Figure 2.8 corresponds to plantation and the right bar to natural regeneration), it becomes clear that for each tree species in plantations negative and positive selection are more frequent than in natural regenerated stands. On the other hand, tree number reductions in natural regenerated stands are more common than in plantations. Independent of the method of regeneration and the tree species, at the European level 75% of all experts recommend positive selections in young stands (51% in German speaking countries). These positive selections are mainly oriented towards the top height of the stands followed by stand age. Other criteria in Germany can be neglected, whereas in Europe approx. one third of the experts cite diameter, tree shape or length of the branch-free bole as criteria for positive selection. Especially for wild cherry diameter of the trees is often used as a decision criterion at the European level. Up to an age of 20 years the number of removed trees per hectare in Europe can be clearly distinguished between plantation and natural regeneration. In planted stands the majority of experts recommend removals of not more than 500

Table 2.9. Frequency of responses on criteria for controlling growth through thinnings.

	Europe			Germany speaking countries		
Criteria:	Ash	Sycamore	Wild cherry	Ash	Sycamore	Wild cherry
Number of trees in the stand	29.3	25.7	21.2	18.5	18.5	9.1
Basal area of the stand	9.8	11.4	9.1	3.7	3.7	0.0
Age of the stand	12.2	14.3	18.2	7.4	7.4	4.5
Top height of the stand	9.8	8.6	9.1	11.1	11.1	13.6
Current total volume increment	4.9	5.7	6.1	0.0	0.0	0.0
Height of FC-trees	2.4	2.9	3.0	0.0	0.0	0.0
Age of FC-trees	0.0	0.0	0.0	0.0	0.0	0.0
DBH of FC-trees	9.8	8.6	12.1	11.1	11.1	13.6
DBH of FC neighboring trees	0.0	0.0	0.0	7.4	7.4	9.1
Other	22.0	22.9	21.2	40.7	40.7	50.0
N*	41	35	33	27	27	22

* Multiple answers or combinations of criteria were allowed.

trees per hectare. In contrast to naturally regenerated stands the number of removed trees varies between 1000 and 4000 trees per hectare. This can be observed for all three tree species.

In Germany the main removals are generally carried out with a maximum number of 500 trees per hectare and the differences between the methods of regeneration are not as distinctive. For planted ash and sycamore 87% of the German speaking experts recommend removals of less than 500 trees per hectare. For naturally regenerated stands this limit is emphasized by 69% of the experts. In planted stands of wild cherry no removals of more than 2000 trees per hectare could be observed. Again, it seems that most of the experts do not differentiate between ash and sycamore, whereas wild cherry is treated specifically.

Thinning

Most of the experts argue for future-crop (FC) tree thinning. Averaging results from both groups questioned and all tree species, only approximately one fifth of all experts recommend other approaches for controlling growth and thinning regimes. Selection based on a certain number of trees per hectare or top height of the stand is considered to be an alternative for deriving thinning intensities and priority. In these cases the main principle for selecting trees to be removed is a free selection thinning with promotion of the best trees in each case (67% and 83% of the European and German speaking experts respectively). Negative selection, i.e. removal of the worst trees, is of secondary importance. As far as future crop trees are selected, the numbers of trees per hectare on average for the European part of the questionnaire are 206 for ash (range from 50 to 500), 189 for sycamore (40 to 400) and 172 for wild cherry (50 to 400). The corresponding numbers for the

German speaking part are 113 for ash (60 to 250), 119 for sycamore (60 to 300) and 96 for wild cherry (15 to 250). By comparing the means and the ranges with the numbers of trees in the final stand, it can be seen that the number of future crop trees is slightly higher than the envisaged number of trees at the end of the production time. Clearly the risk of damages or decreasing vigor of selected trees is counteracted by selecting a reserve pool of trees in younger stand stages. The dominant criteria for defining when the selection of future crop trees should take place in Europe is age (41.2% of the experts) followed by top height of the stand (38.8%) and other criteria such as tree shape, length of the branch-free bole, etc. In Germany age is of little importance; the stage of stand development suitable for future crop tree selection is mainly determined by the length of the branch-free bole (46.4%) followed by top height (44.9%) and age (8.7%). The differences between the tree species are negligible. 44% of the European experts and 54% of the German experts favor permanent marking of the future crop trees.

The most important criteria for controlling growth and deriving thinning intensities in future crop-tree thinnings at the European level is the number of trees in a stand followed by a mixture of other criteria such as length of the branch-free bole, crown development, etc. This can be stated without difference for all three tree species. Although future crop tree thinning mainly focuses on the selected trees, no single tree specific criteria like DBH or age are of any importance for stand treatment in the thinning stage. Nearly all of the experts from Germany who recommend other criteria for controlling thinning mention qualitative criteria to guarantee open-growing conditions from a certain stage of stand development: in other words to make sure that crown base remains at a certain height.

Pruning

Pruning of wild cherry is recommended by 51% of European experts and 79% of the German speaking experts. In both cases the majority prefers regular green pruning between heights of 6 to 10 m. Dry pruning of cherry trees is generally of minor importance. Outside German speaking countries pruning of ash and sycamore is much more common (according to nearly one quarter of the European experts). The distribution of answers pertaining to dry and green pruning is balanced and no specific method of pruning is clearly favored. In German speaking countries pruning of ash is very seldom (only 6.5% of the German experts reported this), 16.1% of the experts recommend pruning sycamore. Again, green pruning is preferred.

2.2.5 Discussion and conclusion

According to experts on valuable broadleaved species common ash, sycamore and wild cherry in Europe will be of higher importance in the future. Reasons for this development are the rising proportions of these species in the total forested area followed by higher proportions in younger age classes and finally a change in the utilization with the goal of increasing the amount of highest paid veneer logs. The opinion of the experts on species and age class proportions base upon national inventories and can be objectified by some examples of published results. In Great

Britain between 1980 and 1998 the proportion of sycamore expands from 48 000 hectares to 61 000 hectares (for ash from 72 500 hectares to 119 000 hectares (Forestry Commission 2003). In Switzerland the proportion of sycamore stands of an age younger than 40 years is 37% and for ash stands 34% (Brassel and Brändli 1999). Comparable inventory results are documented for French (INF 2001) and German (BMELF 1993) forests. In contrast to these verifiable answers of the experts their opinion on the future utilization of the lower ten meters of stem wood is much more subjective. Moreover this opinion reflects an intention more than actual practices and especially for the German speaking part of Europe is driven by concepts presented by Spiecker and Spiecker (1988) and Wilhelm et al. (1999).

The questioning leaves open some doubts whether silvicultural treatments applied to valuable broadleaved tree species in each case are suitable to adapt to the future objective of veneer log production until now. One has to keep in mind that the answers given in the questionnaire with respect to tending and thinning correspond to the recent state of the art and not necessarily to the treatment of the past decades. The result of former forest management is now reflected by the current use of stem wood. This consideration can be proved using wild cherry as an example. Production targets as well as stand treatment are strictly tree species adapted and are common at a European level. Consequently, the amount of high value timber today is much greater than for ash or sycamore. In most cases the experts do not differentiate among these tree species and standard utilization is of lower value.

As an important outcome from this questioning, the development of specific forest management concepts for ash and sycamore as a research need has been identified. These concepts should be able to define consistent production targets with the overall goal of producing high valuable timber while describing the growth controls required to realize these goals based on easily measurable criteria. Corresponding to the economic principle, this maximization of valuable timber must be consistent with the specified costs of the stand treatment.

References

BMELF – Bundesministerium für Ernährung Landwirtschaft und Forsten (ed.) 1990. Bundeswaldinventur 1986–1990. Band I: Inventurbericht und Übersichtstabellen für das Bundesgebiet nach dem Gebietsstand bis zum 03.10.1990, einschließlich Berlin (West). Bonn. 118 p.

Brassel, P. and Brändli, U.-B. 1999. Schweizerisches Landesforstinventar. Ergebnisse der Zweitaufnahme 1993–1995. Haupt. Bern; Stuttgart; Wien. 442 p.

Forestry Commission 2003. National Inventory of Woodlands and Trees – Great Britain. Inventory Report. Edinburgh. 58 p.

INF – Inventaire Forestièr National 2001. Indicators for the sustainable management of French forests. 2000 edition. Ministère de l'agriculture, de l'alimentation de la pêche et des affairs rurales (ed.). Paris. 129 p.

Joyce, P.M.; Huss, J.; McCarthy, R.; Pfeifer, A. and Hendrick, E. 1998. Growing Broadleaves. Silvicultural Guidelines for Ash, Sycamore, Wild Cherry, Beech and Oak in Ireland. COFORD. National Council for Forest Research and Development. Dublin. 144 p.

Kerr, G. and Evans, J. 1993. Growing broadleaves for timber. Forestry Commission Handbook 9. H.M.S.O. London. 95 p.

Roth, E. 1995. Sozialwissenschaftliche Methoden. Lehr- und Handbuch der Forschung und Praxis. 4. durchges. Aufl.; R. Oldenbourg Verlag. München; Wien. 845 p.

Spiecker, M. and Spiecker, H. 1988. Erziehung von Kirschenwertholz. AFZ / Der Wald (20): 562–565.

Spiecker, M. 1994. Wachstum und Erziehung wertvoller Waldkirschen. Mitteilungen der Baden-Württembergischen Forstlichen Versuchs- u. Forschungsanstalt 181. 92 p.

Thies, M. and Hein, S. 2000. Expertenbefragung zur Bedeutung und Bewirtschaftung von Edellaubbäumen im deutschsprachigen Raum. Sektion Ertragskunde im Deutschen Verband Forstlicher Forschungsanstalten. Jahrestagung 2000. Kaiserslautern. 5–7 June 2000. Pp. 256–269.

Thill, A. 1975. Contribution à l'étude du frêne, de l'érable sycomore et du merisier (*Fraxinus excelsior* L., *Acer pseudoplatanus* L., *Prunus avium* L.) Bulletin de la Societé Royale Forestière de Belgique 82: 1–12.

Weber, N. (ed.) 2000. NEWFOR – New Forests for Europe: Afforestation at the turn of the Century. EFI Proceedings 35. European Forest Institute. 244 p.

Wilhelm, G.J., Letter, H.-A. and Eder, W. 1999. Konzeption einer naturnahen Erzeugung von starkem Wertholz. AFZ/ Der Wald 56(5): 232–240.

3. Prerequisites for Growing Valuable Broadleaves

3.1 Genetics and Tree Breeding

Jochen Kleinschmit and Jörg R.G. Kleinschmit
Nordwestdeutsche Forstliche Versuchsanstalt, Göttingen, Germany

Abstract

The importance of genetic quality for timber production is outlined. Human influence on the present genetic structure of valuable broadleaved populations is discussed. The flower biology, natural range, and genetic studies for *Acer pseudoplatanus, Fraxinus excelsior*, and *Prunus avium* are mentioned. Results of provenance experiments for these species are discussed. Methods for implementation of genetic knowledge such as quality seed production in seed orchards or vegetative propagation of superior clones for further improvement are also discussed. Conservation activities for a range of valuable broadleaved species are summarized. The legal situation and recommendations for the use of reproductive material are outlined.

Keywords: genetic situation, provenance experiments, tree breeding, conservation, reproductive material

3.1.1 Introduction

The genetic information and environmental conditions influence the phenotype of a tree. The genotype of a tree determines its adaptability, growth and quality potential. Environmental conditions, including climate, soil and coexisting organisms (e.g. competition), modify the expression of this potential. Depending on the trait, the environmental modification of the phenotype varies (heritability). Phenological traits such as flushing, bud set and length of vegetation period are under strong genetic control and can only slightly be influenced by manipulation of the environment. Other traits such as diameter growth, on the other hand, are considerably modifiable by site conditions, spacing and competition, and thus can be influenced by silviculture. Stands with economically valuable phenotypes can only be grown if both genotype and environment components are favourable.

In forestry the environment can most efficiently, although only slightly, be influenced during the establishment phase. Climate, soil, and coexisting organisms are usually fixed factors where only the silvicultural treatment of the stand can be modified.

The aim of silviculture and tree breeding is to develop phenotypes that serve human needs. Silviculture mainly influences the environment of the trees. To a small extend, depending on the heritability of the traits of interest, the genetic composition of the stand is also altered during thinning operations. Tree breeding, on the other hand, influences the genotypes of the trees by selection, controlled crosses and choice of propagation methods. The environment also interacts with the selected genetic collectives during the selection process.

"Tree breeding is the activity of men to influence and improve forest trees and shrubs with the aim of satisfying human needs as best as possible" (Rohmeder and Schönbach 1959, translation by the authors). The existence of a broad variability of genetically controlled traits is the precondition for tree breeding and thus tree improvement. As a consequence of adaptation to the different ecological conditions within the extended natural ranges of the species, provenance differences are to be expected.

Tree breeding and tree improvement activities include provenance, stand, and single tree testing and selection, grafting and seed orchard establishment, controlled crossing, selection, testing and vegetative propagation of superior clones, and procurement of the material to the forest owner. The different levels of breeding activities (provenance selection, plus tree seed orchards, clonal selection and propagation) allow different levels of improvement which each add approximately 20% above the previous level. This is a rough average from experiments worldwide, which may vary in specific cases considerably. The time span for reliable information from field tests is 10 to 20 years minimum, 10 years for seed orchards from selection until production, and 20 years for a clonal program from the start of selection of superior clones to application. A clonal program needs as a base the existence of provenance/progeny tests.

Provenance, progeny and clonal tests are tools used to describe existing variation and for tree improvement, namely the selection of well performing genetic collections. The tests are established under different ecological conditions to get information about the genetic control of the trait expression and the general adaptability of the genetic units studied. This information cannot be gained from marker studies. Important traits for adaptation include phenological traits, survival, vitality and health. These traits are also a prerequisite for adaptability to heterogeneous site conditions. Beyond these important traits for tree improvement are growth, quality, and wood characteristics.

Provenance experiments with valuable broadleaved species are comparatively new. The establishment of provenance experiments with these species requires different approaches than those of major tree species with extended stands. The collection of a limited number of single tree progenies per population is a compromise to obtain provenance as well as progeny information.

At the time of stand regeneration the forest owner has the unique opportunity to introduce improved material into his forests without considerably increasing the cost of regeneration. Using genetically qualified material for forest regeneration is the most efficient method for the silviculturist to be economically successful.

3.1.2 Genetic situation

In the course of the following text the terms "valuable broadleaved species", "valuable hardwood species" and "noble hardwood species" are used synonymously. These are forest tree species which are not very abundant under natural conditions, are weak competitors and usually grow on rich sites in the lowlands or in the mountain regions. The timber of these species is valuable. Valuable broadleaved species usually occur in mixed hardwood forests, often in small disconnected populations. Some of the species have a high potential for vegetative propagation under natural conditions (e.g. *Prunus avium, Sorbus torminalis*).

Human interference during the past has had drastic implications on the population structure of valuable broadleaved species as the habitats of these species were of special interest for agriculture. Clearing of forests for farming and settlement resulted in a destruction of tree populations, a reduction of population size and the isolation of populations. Furthermore, silviculturists had little interest in these species as compared to more "social" species like oak, beech and conifers. Therefore the abundance of valuable broadleved species was reduced in the remaining forests as well. These conditions often resulted in a narrow genetic base which led to inbreeding effects (loss of vitality), genetic drift (loss of genetic information) and decrease in adaptability.

Plantation of selected cultivars for fruit production or locally maladapted provenances has resulted in introgression of undesirable characteristic traits for wood production into wild (fruit) tree stands.

The following section will focus on *Acer pseudoplatanus* (sycamore), *Fraxinus excelsior* (Common ash) and *Prunus avium* (wild cherry). These are the most common valuable broadleaved species in Central Europe. They have a high economic value and a high potential for afforestation of marginal farmland. There will also be some discussion of rare species such as *Sorbus torminalis* and *Pyrus pyraster*.

Chromosome numbers are n = 26 in *Acer pseudoplatanus* (Spethmann and Namwar 1985, Rohmeder and Schönbach 1959), n = 23 in *Fraxinus excelsior* (Rohmeder and Schönbach 1959), and n = 8 in *Prunus avium* (Kuppers 1979, Rohmeder and Schönbach 1959). All three species are diploid.

Acer pseudoplatanus is insect pollinated and cosexual with functional male or female flowers (Spethmann and Namwar 1985, De Jong 1976).

The natural range of *Acer pseudoplatanus* is concentrated in the Central European mountain range, although small populations can be found at the Baltic Sea. *Acer pseudoplatanus* can be found in the mountains above the beech optimum at an elevational band of 600–800 m and extending up to a maximum elevation of 2000 m in Switzerland (Figure 3.1). Today it is difficult to delineate the natural range of *Acer pseudoplatanus* since it has been extensively planted in the past. As a consequence, the range has extended into the north and east and new genetic informataion has been introduced into some natural populations within the natural range.

Fraxinus excelsior is wind pollinated and can be andromonoecious, gynodioecious, gynomonoecious and trioecious. All forms occur together with hermaphrodite flowers (Hegi 1927). According to Rohmeder (1952) the majority of trees flower either male (39%) or hermaphroditic (41%). Exclusively female flowering trees are rare (4%). Male flowering trees are on average superior in diameter growth due to the fact that female flowers consume more assimilates than male flowers. The natural range of *Fraxinus excelsior* extends to the north as far as Scotland, Denmark and Scandinavia up to the 64° N latitude. The eastern border of

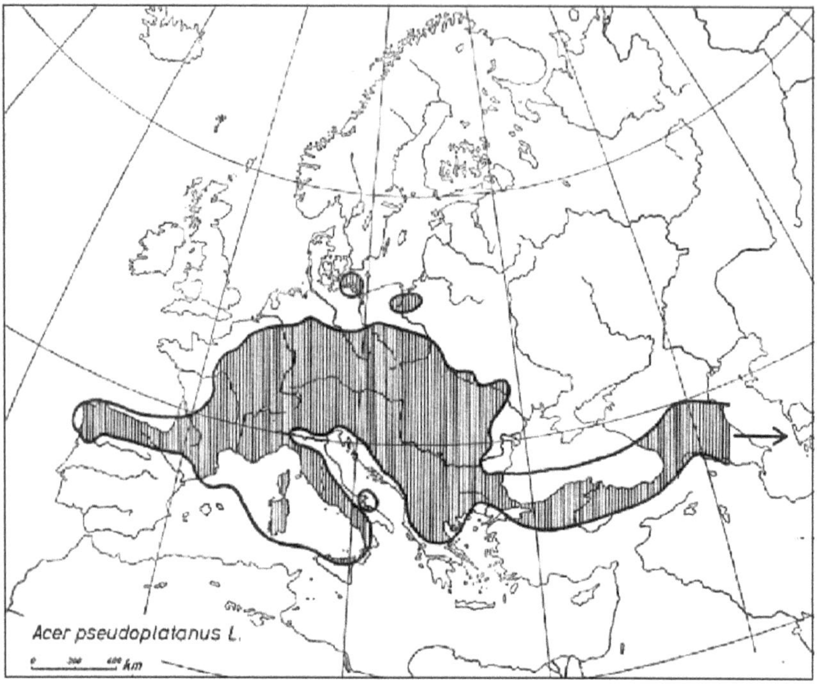

Figure 3.1. Natural range of *Acer pseudoplatanus* (Meusel 1939, modified).

its range is the Wolga river and the southern border extends to Anatolia, the Pindus mountains in Greece, southern Italy and northern Spain (Figure 3.2). *Fraxinus excelsior* grows mainly in low elevations but extends to 700 m in the Harz mountains and 1400 m in the Swiss Alps.

Prunus avium is insect pollinated, hermaphroditic and proterogynous (Hegi 1923). A self incompatibility reproductive system avoids inbreeding (Hilkenbäumer 1953). The natural range of *Prunus avium* covers Europe, Asia Minor and the most northern part of Africa (Krüssmann 1978). In the north of England and Scandinavia it extends up to the 61° N latitude, but is cultivated up to the 67° N latitude (Figure 3.3). The main altitudinal range is between 200 and 600 m, although it can be found in the Central Alps up to 1700 m. *Prunus avium* is regarded as the origin of cultivated forms of sweet cherry and is extensively cross bred with cultivars. This creates problems for forestry since the selection criteria for fruit production (short stem, broad crown, thick branches) are completely different from those for wood production (Kleinschmit et al. 2000).

The genetic situation of a species is described by the use of genetic markers. The marker phenotype allows direct conclusions of the underlying genotype to be made. Genetic markers are a valuable tool for the reconstruction of recolonisation routes, studies of gene flow, relatedness, conservation and identification. The number of known genetic markers is steadily increasing.

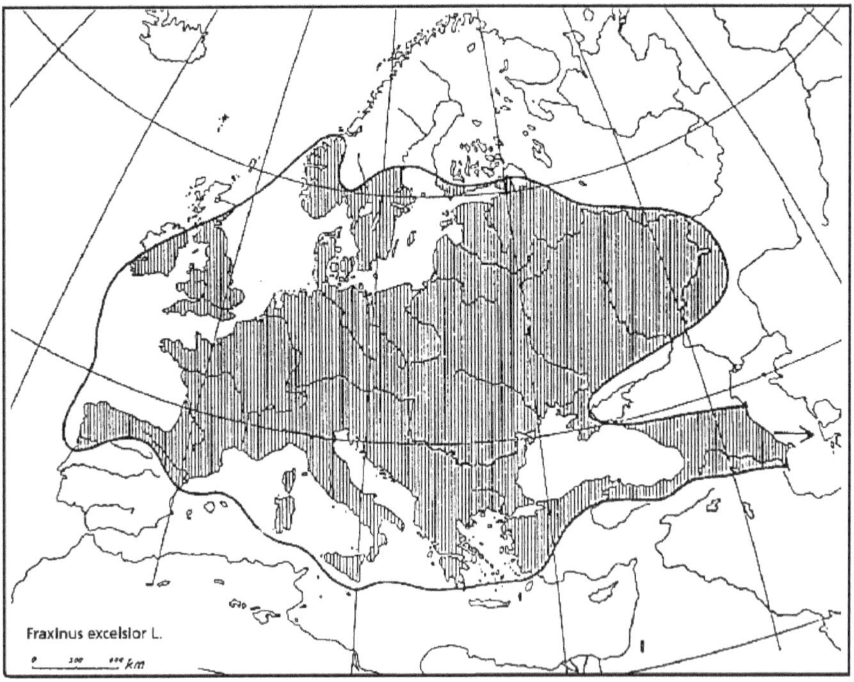

Figure 3.2. Natural range of *Fraxinus excelsior* (Meusel 1939, modified).

Isozyme gene markers were the first tools used to reveal genetic markers. They are easy to use and the biochemical function of the underlying genes can be known. But the number of loci is limited and thus allows only the description of a part of the whole genome's variation to be known.

Theoretically the number of DNA markers is unlimited. But the cost of analysis is high and often the physiological function of the analysed DNA sequence is unknown. For DNA analysis there exist different techniques to analyse either known or anonymous DNA sequences of coding or non coding regions. The DNA markers can be either biparentally inherited via the nucleus or uniparentally inherited via mitochondria and plastids (Gillet 1999). The aim of establishing markers for the expression of quantitative traits (QTL) has not yet been reached for the species under consideration.

The number of studies using gene markers is higher for *Prunus avium*, which is also of interest to horticulturists (e.g. Dirlewanger et al. 2002, Zanetto et al. 2002, Kownatzki 2001, Mohanty et al. 2001, Struss et al. 2001, Dosba et al. 2000), than for the other valuable broadleaved species. Populations of wild cherry are quite differentiated, often unrelated and vegetative propagation occurs regulary. Thus the number of genotypes in these populations is lower than the number of trees (Ducci and Santi 1997). Cultivars of *Prunus* can be genetically differentiated from wild populations by means of genetic markers (Badenes and Parfitt 1995).

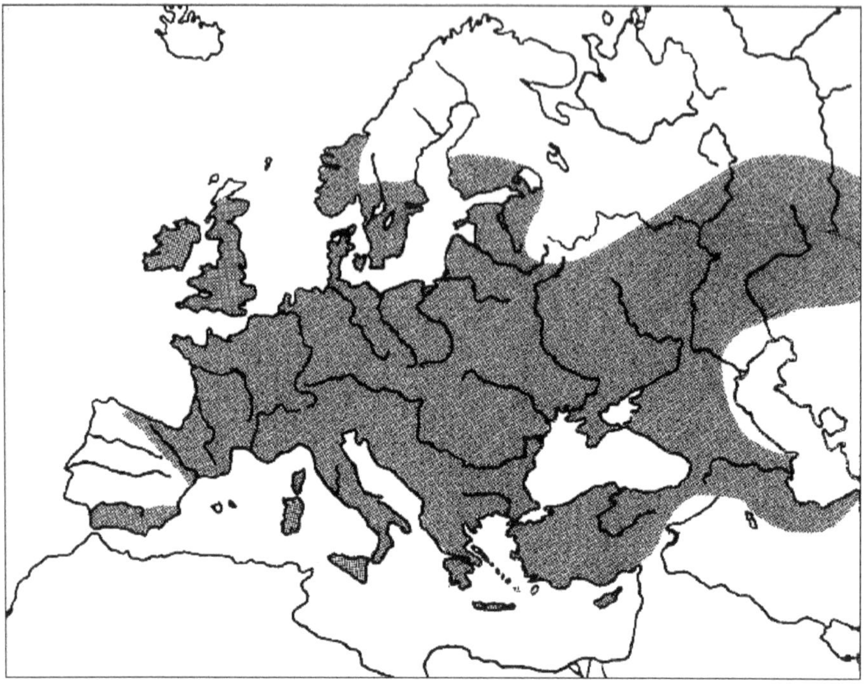

Figure 3.3. Natural range of *Prunus avium* (Meusel 1939, modified).

Isozyme, nuclear, and chloroplast DNA-markers have been established for *Fraxinus excelsior* (Heuertz et al. 2000, Brachet et al. 1999, Lefort et al. 1999) and *Acer* (Bendixen 2000) although only limited information from marker studies is available. For *Fraxinus* an extensive study was started which combined quantitative and genetic traits, as well as propagation methods and utilization aspects, in a project supported by the European Union.

3.1.3 Tree breeding

Provenance differences for *Acer pseudoplatanus* were reported already by Engler (1905) and Schwerin (1907). Provenances from southern Italy did not survive in Germany.

Stands from a limited region in Germany showed strong genetic differentiation especially in stem form in a 15 years old progeny test (Weiser 1981). In Belgium the study of 100 single tree progenies, originating from 20 Belgium stands, discovered regional differences in seed size (Galoux and Falkenhagen 1965). Growth potential decreases with increasing elevation and increases with increasing temperature from the site of origin (Galoux 1969).

Seeds for an *Acer pseudoplatanus* provenance experiment including 45 provenances, each represented by 5 single trees, from Germany, Switzerland, Austria, Czechoslovakia, Yugoslavia, Romania and Denmark were collected in 1981 by the Lower Saxony Forest Research Institute and planted on 10 field experiments in northern and western Germany. At age 16 of the field experiments there existed pronounced ecotypic patterns of variation. Some local provenances as well as some provenances from elsewhere showed very good performance at the test sites. Some of the variation in phenological characters can be explained by climatic differences of the site of origin (clinal variation). For other characters no clear geographic pattern can be found. In height growth and stem form the highest performing provenances originated from Rosengarten (near Hamburg), Bodenmais in Bavaria, and Prachtice and Karlovice in the Czech Republic. Provenances from Romania were average. The fastest and slowest growing provenances have a height 30% above and 15% below respective means.

Since *Fraxinus excelsior* occurs on dry limestone sites as well as in alluvial plains, the existence of two races has been supposed by Münch and Dieterich (1925). The studies of Weiser (1995, 1974, 1964, 1962) and Schönborn (1967), however, did not prove any genetic differences between these types. This means that sufficient gene exchange between these units has occurred and/or that the selection pressure from soil conditions is not as strong as the selection pressure due to climatic differences between geographic regions.

Seeds for a *Fraxinus excelsior* provenance experiment including mainly German sources (43), in addition to some from Switzerland (3), Austria (4) and Romania (2) each represented by 5 single trees, were collected in 1982 and sown in 1985 by the Lower Saxony Forest Research Institute. Twenty-six field experiments have been established in Germany (19), France (2), Belgium (2), The Netherlands (2) and Romania (1). The first results have been published by Kleinschmit et al. in 1996 and 2001. Under German conditions Romanian provenances flushed early and were damaged by late frost and therefore had the worst stem form. Phenological characteristics showed a clinal pattern of variation. For height growth and stem form the pattern of variation displayed strong ecotypic components, indicating considerable differences between populations within a short distance. Variation of trait expression within provenances was as high as variation between provenances.

For *Prunus avium* two limited provenance studies including German, French, Yugoslavian and Romanian material were established by the Lower Saxony Forest Research Institute in 1985. Unpublished results show that variation within provenances is as high as between provenances at age 15. French provenances have similar growth as the German provenances. Those from Yugoslavia and Romania are inferior under German growing conditions. There exist drastic differences in stem form between and within progenies, thereby offering an excellent potential for selection of superior clones for vegetative propagation.

The results of the aforementioned provenance experiments show that the pattern of variation is different for different traits. Phenological traits, which are under strict genetic control, show a clinal pattern of variation. This can be explained by a direct response to natural selection pressure. Growth traits on the other hand show more ecotypic patterns, indicating that the selection of tested stands can already lead to a considerable improvement. The results of provenance experiments stress the importance of choosing carefully the origin of material for a specific plantation site.

Propagation by grafts is possible with *Acer pseudoplatanus, Fraxinus excelsior*, and *Prunus avium*. In Sycamore special wood structures occur such as "curly maple" and "bird's eye maple". Genotypes showing these trait expressions can be vegetatively propagated by grafts; the grafts preserve these special characteristics.

Only for *Prunus avium* do methods of cutting propagation have a sufficient rooting success and methods of in vitro propagation are developed for commercial use (Meier-Dinkel 2001, 2000). For *Prunus avium* superior clones have been selected in various countries and reproduced vegetatively by in vitro propagation (Meier-Dinkel et al. 1997). Such clones can be used in clonal mixtures of 30 or more, thereby offering an excellent base for the production of valuable timber.

For *Fraxinus excelsior* first results with in vitro propagation are available although they must be developed further (Hammatt 1996, Silveira and Cottignies 1993).

Grafted seed orchards have been established on a regional basis with selected superior adult trees for all the species (Table 3.2). Seed orchards can be used to overcome the problems of coancestry, inbreeding, and genetic drift, which are present in many of the small populations. They can also be used to select for stem form. Thus seed orchard seeds are the most interesting base material for silviculture with valuable broadleaved species. Seed orchards of *Prunus avium* are currently included in a joint testing program participated by different European countries (Germany, The Netherlands, Belgium). When test results are available, the seed material from tested seed orchards should be preferred over seeds from untested sources.

3.1.4 Genetic conservation

The aim of genetic conservation of forest tree species is to maintain the natural patterns of variation and the adaptability of the species to future environmental changes in particular. Today the natural patterns of variation rarely exist in the valuable broadleaved species. The long human impact on forest surface and composition has changed the natural pattern drastically. The implication of these changes is poorly understood.

Over the long term genetic conservation of forest tree species in a densely populated country like Germany can only be successful if the species is used in a regular forest management system. This approach is practicable with sycamore, Common ash and wild cherry. In the past, however, the integration of conservation and utilization was a problem with most of the minor hardwood species due to their rareness. Often reproductive units no longer existed. Even in the case of sycamore, Common ash and wild cherry, the integration of conservation and utilization will be more successful if the economic return of the silviculture with these species is of interest to the forest owner. This will be the case if the value of the timber is high. Tree breeding and conservation have an important link. All activities which improve the economic return will directly influence the participation of the species in artificial regeneration programs. The increasing timber prices for noble hardwood during the last decade had a direct influence on the request for plant material of noble hardwood species. This provides support for the conservation of these species.

In Europe the programs for conservation of forest tree species are coordinated by EUFORGEN (European Forest Genetic Resources Programme) which is located at the International Plant Genetic Resources Institute in Rome, Italy. There exists a special network for noble hardwood species that was established in 1996 (Turok et al. 1996.). It has been made clear that most neighboring countries have an interest in the same species. Since the ecological conditions do not necessarily respect political boundaries, it is sensible to coordinate the genetic conservation programs on a European level.

Conservation can be in situ with stands or single trees or ex situ with plantations, seed orchards, clonal archives and storage of seed, pollen, tissue and even DNA sequences. In situ conservation stands have the advantage of existing within current evolutionary processes and being integrated into normal forest management systems. A management restriction is that regeneration has to be either natural or of material from the same stand. In situ conservation in stands is only meaningful if the populations are large enough. In situ conservation of single trees can only be an intermediate step because they will die naturally at some time. Thus single trees have to be conserved ex situ.

Ex situ conservation in plantations, seed orchards and clonal archives is relatively easy. The choice of the ex situ conservation site should allow evolutionary adaptational processes to take place. Seed orchards for ex situ conservation facilitate the creation of new, genetically diverse populations of unrelated individuals. These seed orchards may represent a certain level of genetic improvement as well. If in situ conservation measures are taken, the improvement aspect of ex situ conservation activities can be stressed without the risk of loosing important genetic information. This is the case with sycamore, Common ash and wild cherry.

Seed orchards have the additional advantage that reproductive material can be harvested soon. Storage of seed, pollen, tissue and DNA sequences must be done under controlled environmental conditions. These conservation measures are technically susceptible and become increasingly sophisticated from beginning to the end of the enumeration.

Tree breeding includes many conservation activities as well. Therefore a broad overlap between tree breeding and conservation of forest genetic resources exists so long as breeding activities succeed in maintaining the important genetic diversity.

The general conservation situation of noble hardwood species in Germany is summarized in Table 3.1.

Acer pseudoplatanus and *Fraxinus excelsior* are species which still have sufficiently large populations. The trade of reproductive material is controlled under the restriction of EC legislation (Council directive 1999). Therefore the conservation strategy can include an important in situ conservation component. This is not true to the same extent for *Prunus avium* where small population sizes are frequent, where hybridization with cultivars occurs, and where no control of the reproductive material has existed until now. Seed originating from countries with different ecological conditions was used as well as seed from jam factories. Problems with small population size and hybridization, which are typical for all

Table 3.1. Common species of valuable broadleaves in Germany (from Kleinschmit et al. 1996, modified) (number of stars indicate importance of the factor on the species).

			Reason of endangerment			
Species	Importance	Species under council directive 1999/105/EC	Hybridisation with other species	Hybridisation with cultivars	Small population size in breeding, genetic drift	Diseases
Native						
Acer campestre	**				*	
Acer platanoides	***	*			**	
Acer pseudoplatanus	*****	*				
Alnus glutinosa	****	*				
Fraxinus excelsior	*****	*				
Malus sylvestris	*			***	***	
Prunus avium	****	*		*	*	*
Pyrus pyraster	*			***	***	
Sorbus domestica	*			*	**	
Sorbus torminalis	**				*	
Tilia cordata	***	*	*			
Tilia platyphyllos	**	*	*		*	
Ulmus glabra	***		*		*	***
Ulmus laevis	**		*		*	***
Ulmus minor	**		*		*	****
Non native						
Juglans regia	*				*	
Juglans nigra	*				*	

wild fruit species, can partly be handled by ex situ conservation in seed orchards and clonal archives. The problems are much more pronounced with some of the other valuable broadleaved species where in situ conservation possibilities are extremely restricted. Here the main emphasis has to be on ex situ conservation. This is especially true for the other wild fruit species (*Malus sylvestris* and *Pyrus pyraster*). In *Ulmus* the problems with Dutch elm disease restrict even ex situ conservation methods since seed orchards will also be attacked sooner or later by the disease. Clonal archives, which are cut back regularly, allow good sanitary conditions to be maintained.

The main emphasis of conservation of forest trees in Germany is on the management of in situ conservation stands. With the valuable broadleaved species the activities shift more in the direction of ex situ conservation. The status of activities in Germany up to the year 2000 is summarized in Table 3.2. From this table it is obvious that only few of the species, *Acer pseudoplatanus, Alnus glutinosa, Fraxinus excelsior*, and *Tilia cordata* still have stands of reasonable size. For those in stands that are rare or very small, the main emphasis is on single trees in situ which are entered into clonal archives or seed orchards. For more details of other species and activities in other countries, the publications of the Noble Hardwood Network (Turok et al. 1999, 1998) are a valuable source of information.

3.1.5 Legal situation

Until recently the trade of reproductive material of most valuable broadleaved species was not under legal control. Exceptions were *Fraxinus excelsior* and *Acer pseudoplatanus* in Germany. With the new EU regulations (Council directive 1999) this situation will change and more species will be included in the national laws. The past situation favoured the silvicultural use of reproductive material originating from low wage countries for regeneration, irrespective of the ecological conditions at the site of origin. This can have implications for the existing populations where uncontrolled mixing of provenances has occurred. The non-adapted material represents a genetic load for the population when flowering begins. Pollen and seed contamination will increase with an increasing percentage of non-adapted plants.

In Germany the "Deutsche Kontrollvereinigung" (German Control Organisation) was established by private seed dealers, nurseries, forest owners, and public forest services to guarantee the control of reproductive material of high quality selected stands. This organisation circumvents at least part of the disadvantages due to the import of non-adapted seed linked to the non-existing legal administrative control of seed transfer.

3.1.6 Recommendations for the use of reproductive material

Natural regeneration of forest stands is the present preference in many countries. This should be critically checked for genetic diversity and genetic quality of the existing stands. It is not worthwhile to naturally regenerate every population from

Table 3.2. Conservation of valuable broadleaved species in Germany up to the year 2000.

	in situ			ex situ					
species	stands		single trees	stands		seed orchards			clonal archives
	number	area (ha)	number	number		number	area (ha)	number of fam./clones	number clones
Acer campestre	12	5	1498	6	0,4	1	2	80	79
Acer platanoides	26	7	872			3	5	165	6
Acer pseudoplatanus	127	155	498	6	10	17	28	995	181
Alnus glutinosa	139	341	227			21	37	1002	32
Fraxinus excelsior	156	286	188			11	24	719	143
Malus sylvestris	3	2	3262	38	8	23	22	739	78
Prunus avium	48	26	1184	11	8	20	38	1270	276
Pyrus communis	17	8	1859	17	4	20	14	628	66
Sorbus aucuparia	11	5	391	15	6	4	4	286	
Sorbus torminalis	253	19	2015	13	6	8	9	434	46
Tilia cordata	148	148	260			20	42	999	159
Tilia platyphyllos	129	22	599			5	10	538	192
Ulmus glabra	123	62	3574	7	7	9	13	352	843
Ulmus laevis	124	67	3844	14	5	4	6	370	250
Ulmus minor	41	12	880			2	5	105	143
Total	1357	1165	21151	127	54	168	259	8682	2494

an economic point of view. Criteria must include the population size, growth and quality of the existing stand. Only if the population size is at least 20 individuals of reproductive age and growth and quality are above average is natural regeneration recommended (Melchior et al. 1989). Otherwise enrichment planting with improved material or, in extreme situations, replacement of undesirable stands should be considered.

For practical seed and plant procurement there should be a clear ranking of the base material according to its status of genetic improvement:

1. Mixtures of tested clones; vegetatively propagated (available for *Prunus avium*);
2. Seed orchard material from tested seed orchards;
3. Seeds from tested stands (available for some valuable broadleaved species);
4. Seed orchard material (available for all valuable broadleaved species) and
5. Selected phenotypically superior stands (available for all species under legal control).

The forest owner should request the material from the nurseries in this sequence (1. being the highest priority) and never plant material which is not genetically qualified as above average.

The seed orchards established on a regional base in Germany can be seen in Table 3.2. In Germany for *Acer pseudoplatanus*, *Fraxinus excelsior* and *Prunus avium,* seventeen, eleven and twenty seed orchards exist respectively. When these are in full production, they will produce sufficient seed to supply the majority of potential plantations in Germany.

As long as the seed orchards are not tested, preference should be given to the seed orchards which correspond regionally to the plantation zone. However, due to the results of provenance testing, these zones must not be very fine grained. In Germany it seems to be sufficient for valuable broadleaved species to differentiate between northern lowlands, mountain region and alpine region. Eventually these regions can be subdivided into west and east.

Different European countries and the German states procure provenance recommendations which follow our line of discussion (e.g. Centrum voor Genetische Bronnen, Nederland 2002, Baden-Württemberg Ministerium ländlicher Raum 1997, Sächsische Landesanstalt für Forsten 1996, Niedersächsisches Ministerium für Ernährung, Landwirtschaft und Forsten 1995, Bayerisches Staatsministerium für Ernährung, Landwirtschaft und Forsten 1987).

3.1.7 Conclusions

The appreciation of what we call "valuable hardwood species" today has changed considerably during forest history. A lack of interest of these species during the past centuries in forest management has had severe influence on their gene pools. Local extinction, fragmentation, pollination by cultivars and unregulated seed transfer changed the original pattern of variation considerably and resulted in quite an unpredictable local genetic situation. This situation should be taken into account by forest owners and forest managers when selecting reproductive material for these species. The efforts of genetic conservation and tree breeding for these species during the last 50 years can help to overcome some of the obvious mistakes created by unregulated human interferences during the previous centuries, and at the same time help to improve stability, health and productivity of these increasingly economically interesting species.

Acknowledgements

We thank Ulrike Seifert and Heidrun Terasa, Dept. Forest Genetic Resources, Lower Saxony Forest Research Institute, for technical assistance.

References

Badenes, M.L.and Parfitt, D.E. 1995. Phylogenetic relationships of cultivated *Prunus* species from an analysis of chloroplast DNA variation. Theor. Appl. Genet. 90: 1035–1041.

Baden-Württemberg Ministerium Ländlicher Raum 1997. Empfohlenen Herkünfte forstlichen Vermehrungsgutes für Baden-Württemberg. Landesforstverwaltung Baden Wüttemberg.

Bayerisches Staatsministerium für Ernährung, Landwirtschaft und Forsten 1987. Empfohlene Herkünfte forstlichen Vermehrungsgutes für Bayern.

Centrum voor Genetische Bronnen, Nederland, 2002 7[e] Rassenlijst van Bomen. Stichting DLO, Centrum voor Genetische Bronnen, Nederland.

Bendixen, K. 2000. Reproductive system of field maple (*Acer campestre* L.). Dissertation, University Göttingen.

Brachet, S., Jubier, M.F., Richard, M., Jung-Muller, B. and Frascaria-Lacoste, N. 1999. Rapid identification of microsattelite loci using 5′anchored PCR in the common ash *Fraxinus excelsior*. Molecular Ecology 8: 160–163.

Council directive 1999. On the marketing of forest reproductive material. 1999/105/EC.

De Jong, P. C. 1976. Flowering and sex expression in *Acer* L. A biosystematic study. Mededelingen, Landbouwhogeschool Wageningen, Nederland.

Dirlewanger, E., Cosson, P., Tavaud, M., Aranzana, M.J., Poizat, C., Zanetto, A., Arús, P. and Laigret, F. 2002. Development of microsatellite markers in peach [*Prunus persica* (L.) Batsch] and their use in diversity analysis in peach and sweet cherry (*Prunus avium* L.). Theor. Appl. Genet. 105: 127–138.

Dosba, F., Fischer, M. and Zanetto, A. 2000. 1. Genetic Resources – International Network on Prunus Genetic Resources (GEN RES 'EU 061'): Acta horticulturae. The Hague.. Pp. 113–116.

Ducci, F. and Santi, F. 1997. The distribution of clones in managed and unmanaged populations of wild cherry. Can. Jour. For. Res. 27: 1998–2004.

Engler, A. 1905. Einfluss der Provenienz des Samens auf die Eigenschaften der forstlichen Holzgewächse. Bergahorn (*Acer pseudoplatanus* L.). Mitteilungen der Schweizerischen Zentralanstalt für das Forstliche Versuchswesen 8: 225–236.

Galoux, A. 1969. Diversification génécologique régionale chez les espèces ligneuses feuillis. Oecologia Plantarum 4: 1–14.

Galoux, A. and Falkenhagen, E. 1965. La samare, les correlations et la croissance au stade infantile. Recherches sur la variabilité génécologique de l´erable sycomore (*Acer pseudoplatanus* L.) en Belgique. Station de Recherches des Eaux et Forêt Groenendaal – Hoeilaart 10. 60 p.

Gillet, E.M. (ed.) 1999. Which DNA Marker for Which Purpose? Final Compendium of the Research Project Development, optimisation and validation of molecular tools for assessment of biodiversity in forest trees. European Union DGXII Biotechnology FW IV Research Programme Molecular Tools for Diversity, Online book: http://webdoc.sub.gwdg.de/ebook/y/1999/whichmarker

Hammatt, N. 1996. *Fraxinus excelsior* L. (Common Ash). In: Bajaj, Y.P.S. (ed.), Biotechnology in Agriculture and Forestry, Trees IV. Pp. 173–193.

Hegi, G. 1923. Illustrierte Flora von Mitteleuropa. Lehmann Verlag München.

Hegi, G. 1927. Illustrierte Flora von Mitteleuropa. Carl Hanser Verl. München.

Heuertz, M., Hausman, J.-F., Tsvetkov, I., Frascaria-Lacoste, N. and Vekemans, X. 2000. Assessment of genetic structure within and among Bulgarian populations of the common ash (*Fraxinus excelsior* L.). Molecular ecology 10(7): 1615–1624.

Hilkenbäumer, A. 1953. Obstbau. Grundlagen, Anbau und Betrieb. Dritte Auflage. Paul Parey Verlag, Berlin – Hamburg.

Kleinschmit, J., Svolba, J., Enescu, V., Franke, A., Rau, H.-M. and Ruetz, W. 1996. Erste Ergebnisse des Eschen – Herkunftsversuches von 1982. Forstarchiv 67: 114–122.

Kleinschmit, J., Stephan, B.R., Liesebach, M., Schüte, G. and Steinhoff, S. 1996. Noble Hardwood species in Germany: occurrence and gene conservation measures. In: Turok, J., Eriksson, G., Kleinschmit, J. and Canger, S. (eds.), Noble Hardwoods Network, Report of the first meeting 24–27 March 1996, Escherode, Germany. International Plant Genetics Resources Institute, Rome, Italy. Pp. 101–110.

Kleinschmit, J., Spellmann, H., Rumpf, H., Guericke, M. and Wachter, H. 2000. Entscheidungshilfen zur Bewirtschaftung der Vogelkirsche in Nordwestdeutschland. Forst und Holz 55: 611–616.

Kleinschmit, J., Lück, F.W., Rau, H.-M. and Ruetz, W. F. 2001. Ergebnisse eines Eschen-Herkunftsversuches. (Results of an Ash Provenance Experiment). Forst und Holz 57: 166–172.

Kownatzki, D. 2001. Asexual and sexual reproduction in populations of wild cherry (*Prunus avium*). Dissertation, University Göttingen.

Krüssmann, G. 1978. Handbuch der Laubgehölze. Paul Parey Verl. Berlin – Hamburg.

Kuppers, H. 1979. Problematik der Veredelungsunterlagen für Sauer- und Süsskirschen im Spiegel von 250 Jahren. *Prunus avium* L. Vogelkirsche. Deutsche Baumschule 30: 21–27.

Lefort, F., Brachet, S., Frascaria-Lacoste, N., Edwards, K.J. and Douglas, G.C. 1999. Identification and characterisation of microsattelite loci from ash (*Fraxinus excelsior* L.) and their conservation in the Oleaceae family. Molecular Ecology 8: 1075–1092.

Meier-Dinkel, A. 2000. Untersuchungen zur Kommerzialisierung der Mikrovermehrung von *Prunus avium* im Rahmen eines BMBF-Verbundprojektes. Tagung der deutschen Sektion der International Association for Plant Tissue Culture (IAPTC), Bonn. 5–7.10.2000. Abstracts.

Meier-Dinkel, A., Svolba, J. und Kleinschmit, J. 1997. Selektierte, mikrovermehrte Vogelkirschen-Klone. AFZ/Der Wald 18: 963–964.

Meier-Dinkel, A. 2001. In vitro Vermehrung von Vogelkirsche. In: Jahresbericht der Niedersächsischen Forstlichen Versuchsanstalt.

Melchior, G.H., Becker, A., Behm, A., Dörflinger, H., Kleinschmit, J., Muhs, H.J., Schmitt, H.-P., Stephan, B.R., Tabel, U., Weisgerber, H. and Widmaier, T. 1989. Konzept zur Erhaltung forstlicher Genressourcen in der Bundesrepublik Deutschland. Forst und Holz 44: 379–404.

Meusel, H. 1939. Verbreitungskarten mitteleuropäischer Leitpflanzen. 3. Reihe, Herzynia 2, Halle.

Mohanty, A., Martín, J.P. and Aguinagalde, I. 2001. A population genetic analysis of chloroplast DNA in wild populations of *Prunus avium* L. in Europe. Heredity 87(4):421–427.

Münch, E. von and Dieterich, V. 1925. Kalk- und Wassereschen. Silva 13: 129–135.

Niedersächsisches Ministerium für Ernährung, Landwirtschaft und Forsten, 1995. In: Abt., C (ed.), Empfohlene Herkünfte forstlichen Vermehrungsgutes für den Anbau in Niedersachsen und Schleswig-Holstein. Niedersächsische Forstliche Versuchsanstalt.

Rohmeder, E. 1952. Untersuchungen über die Verteilung der Geschlechter bei den Blüten von *Fraxinus excelsior*. Forstwiss. Centralbl. 71: 17–29.

Rohmeder, E. and Schönbach, H. 1959. Genetik und Züchtung der Waldbäume. Paul Parey Verl. Hamburg, Berlin.

Sächsische Landesanstalt für Forsten, 1996. Herkunftsgebiete und Herkunftsempfehlungen für Forstliches Saat- und Pflanzgut im Freistaat Sachsen. Schriftenreihe der Sächsischen Landesanstalt für Forsten, 10/96.

Schönborn, A. von 1967. Gibt es Bodenrassen bei Waldbäumen? Allgem. Forstzeitschrift 22: 294–296.

Schwerin, F. Graf von, 1907. Über Pflanzeneinbürgerung sowie Gründe und Abwendung vorkommenden Misslingens. Mittelungen der Deutschen Dendrologischen Gesellschaft 16: 148–157.

Silveira, C.E. and Cottignies, A. 1993. Period of harvest, sprouting ability of cuttings, and in vitro plant regeneration in *Fraxinus excelsior*. Canadian Journal of Botany 72: 261–267.

Spethmann, W. and Namwar, K. 1985. Der Bergahorn und die Gattung *Acer*. Allgemeine Forstzeitschrift 42.

Struss, D., Boritzki, M., Glozer, K. and Southwick, S.M. 2001. Detection of genetic diversity among populations of sweet cherry (*Prunus avium* L.) by AFLPs. The journal of horticultural science 76(3): 362–367.

Turok, J., Eriksson, G., Kleinschmit, J. and Canger, S. 1996. Noble Hardwoods Network. International Plant Genetic Resources Institute. 172 p.

Turok, J., Collin, E., Demesure, B., Eriksson, G., Kleinschmit, J., Rusanen, M. and Stephan, R. 1998. Noble Hardwoods Network: Second Meeting. International Plant Genetic Resources Institute. 103 p.

Turok, J., Jensen, J., Palmberg-Lerche, Ch., Rusanen, M., Russell, K., de Vries, S. and Lipman, E. 1999. Noble Hardwoods Network: Third Meeting. International Plant Genetic Resources Institute. 116 p.

Weiser, F. 1962. Beitrag zur Frage der rassischen Differenzierung von Esche (*Fraxinus excelsior* L.) trockener Kalkstandorte und grundwasserbeeinflusster Standorte. Dissertation Gaterslenben.

Weiser, F. 1964. Anlage und erste Ergebnisse vergleichender Anbauversuche mit generativen Nachkommenschaften von Eschen (*Fraxinus excelsior* L.) trockener Kalk- und grundwasserbeeinflusster Standorte. Forstwiss, Centralbl. 83: 193–211.

Weiser, F. 1974. Ergebnisse 10-jähriger vergleichender Anbauversuche mit generativen Nachkommenschaften von Eschen (*Fraxinus excelsior* L.) trockener Kank- und grundwasserbeeinflusster Standorte. Beitr. für die Forstwirtschaft 8: 11–16.

Weiser, F. 1981. Zielstellung und Ergebnisse einer 15-jährigen Bestandesnachkommenschaftsprüfung bei Bergahorn (*Acer pseudoplatanus* L.). Beiträge für die Forstwirtschaft 15: 142–144.

Weiser, F. 1995. Beitrag zur Existenz von Ökotypen bei gemeiner Esche (*Fraxinus excelsior* L.). Forstarchiv 66: 251–257.

Zanetto, A., Maggioni, L., Tobutt, K.R. and Dosba, F. 2002. *Prunus* genetic resources in Europe: Achievement and perspectives of a networking activity. Genetic resources and crop evolution 49(3): 331–338.

3.2 Diseases, Disorders and Pests of Selected Valuable Broadleaved Tree Species

Alfred Wulf and Rolf Kehr
Institute for Plant Protection in Forests, Federal Biological Research Centre for Agriculture and Forestry, Braunschweig, Germany

Abstract

Pests, diseases and other disorders can pose life-threatening problems for valuable broadleaved tree species and effectively prevent their economic success on certain sites or under certain circumstances. Potentially damaging factors must be identified before making the decision to plant or cultivate a certain tree species, be it in forests or in plantations. In many cases control possibilities for diseases and pests are very limited, especially those methods of chemical control which have been the focus of much research in the past. However, knowledge of predisposition, genetic qualities of the host tree and factors in disease and pest development can help reduce risk and select suitable tree species for a given site. This chapter gives a short treatise of important disorders affecting some of the most valuable broadleaved tree species in Europe.

Keywords: pests, diseases, valuable broadleaved species, predisposition, genetic qualities

3.2.1 Preface

In the context of a short chapter it is not possible to provide a comprehensive treatise of the diseases, pests and abiotic disorders of the large group of trees termed "broadleaved". Although stands of beech and oak can be very valuable, these common tree species are not treated here, but rather the focus is on species where the individual tree can attain high value when proper silvicultural methods are employed. Only those problems will be presented which are either very common, rather conspicuous or pose a certain economic threat due to their damage potential. The selection is of course subjective and influenced by the experience and interests of the authors. Also, only those disease-causing organisms which are more or less host specific have been taken into consideration. For instance, the wide range of macromycetes causing an unspecific wood rot in advanced life stages of trees has mostly been disregarded. For this group, which is of special importance for safety aspects of amenity trees, a large selection of literature is available (e.g. Schwarze et al. 1999, Butin 1995, Breitenbach and Kränzlin 1986). There is also a wealth of other diagnostic literature available for reliable identification of many disease-causing organisms on trees (Strouts and Winter 2000, Brandenburger 1985, Ellis and Ellis 1985, Browne 1968). Damage caused by mammals, such as browsing and bark stripping by Red deer (*Cervus elaphus*), is usually not specific to individual tree species and is therefore also not treated. The same holds true for polyphagous insect species such as common cockchafer (*Melolontha melolontha*).

Some of the pest organisms treated in this chapter are not native to Europe and have either been recently introduced or could pose a real threat to wood production if they were introduced. Such species usually possess high damage potential since they have not co-developed with their new host. Even though the possible introduction and the ensuing damage are difficult to predict, two exemplary organisms which could cause serious damage to valuable broadleaved species (Sirococcus canker of walnut and Asian longhorned beetle) will be discussed here.

3.2.2 Maple

In the case of maples, from a forestry point of view, there are few disease-causing organisms which pose a lethal threat to their host. However, there is a wide range of conspicuous leaf diseases and disorders. With the exception of gall-inducing organisms, maple species seem to be more attractive to fungi than to insects.

Abiotic influences

Of the maple species cultivated in Europe, *Acer pseudoplatanus* and *A. platanoides* are the most important and possess the highest value for forest production. Young trees are somewhat prone to late spring frosts, but otherwise they naturally regenerate and grow well on suitable sites. Older trees are known to develop frost damage and sun scald on the south-facing side of the tree following cold winters. Both species have a superficial periderm as does beech at a younger age and are thus susceptible to high bark temperatures. In *A. pseudoplatanus* this problem is gradually resolved at later stages of maturity when rhytidome formation sets in.

Coral spot

The causal agent of coral spot, *Nectria cinnabarina* (Tode) Fr., is a pathogen of worldwide distribution which also occurs as a saprophyte on many plant species, especially broadleaved trees. It can cause extensive necrosis of bark and cambium as a weak parasite when host predisposition allows for it and is therefore of phytopathological interest. This fungus is discussed in detail in many standard textbooks on pathology of woody plants because of its broad host range (e.g. Butin 1995, Sinclair et al. 1993, Jahn 1990). On forest tree species it is most commonly found on *Tilia, Ulmus* and *Carpinus* and especially on *Acer* where it causes the greatest damage.

Symptoms are often seen in early summer when shoots above fresh bark necroses start to die. If the cambial damage encircles the whole main shoot, it dies completely above the necrotic lesions; a phenomenon encountered in younger trees especially. Reliable macroscopic identification of the causal organism is possible only after the growing season when the characteristic fungal fruit bodies develop on dead bark. These bright red, cushion-shaped, several mm large pustules have led to the name "coral spot". They are the sporodochia of the fungal anamorph (*Tubercularia vulgaris*) which develop on a subperidermal stroma and form hyaline, ellipitical to cylindrical conidia measuring 3 × 6 µm. Towards the end of winter the 0.5 mm large, cherry-red perithecia of the *Nectria cinnabarina* teleomorph are formed in groups alongside the gradually shrinking sporodochia, thus producing evenly two-celled ascospores measuring 6 × 16 µm.

This fungus acts as a weak parasite and wound parasite especially when the host plant is predisposed which can be the case following severe frost, wounding or, as in most cases, water deficit, which can lead to extensive *Nectria* damage. Water deficit is caused in particular by a loss of root mass and longer dry periods associated with the transplanting of nursery trees. Therefore this disease is one of the more serious problems affecting freshly planted city and landscape trees as extensive damage can be encountered in trees suffering from prolonged periods of stress associated with planting ("planting shock"). Infections can also be caused by pruning measures on amenity trees. Correct pruning techniques and, in some cases (e.g. in orchard and plantation trees), the use of effective wound dressings can help reduce the negative effects of branch pruning (Riecken 1985).

In forests *Nectria cinnabarina* is able to cause damage primarily to young maple trees that have been planted for forest production and, to a lesser extent, to stands arising from natural regeneration. Incorrect planting procedures using plants with poorly developed root systems, low water retention capacity of the soil or prolonged drought periods following planting are possible causes for a high incidence of coral spot. In recent years dieback of older *Acer pseudoplatanus* trees and serious damage to older forest stands of this species has been reported in Germany; in many cases *Nectria cinnabarina* as well as *N. coocinea* and, in some cases, *N. ditissima* are apparently associated with the symptoms. Climatic extremes are held responsible for the disease development, especially long periods of extreme frost in late autumn when the bark is not yet physiologically adapted to such low temperatures (Bressem 2002).

Sooty bark disease

Sooty bark disease is a bark disorder associated with the fungus *Cryptostroma corticale* (Ell. & Ev.) Gregory and Waller, which is known as a saprophyte from North America and Europe but has occasionally been implicated in active bark and cambium damage of *Acer pseudoplatanus* and *A. platanoides* in Great Britain, France and several other European countries (Butin 1995, Peace 1962). Symptoms consist of death of inner bark and cambium followed by bark blistering and shedding. As the bark falls off it exposes the black spore masses of the fungus which are formed in stromatic layers within the bark tissue. Conidia are pale brown when ripe and measure

4–6 × 3.5–4 µm. Depending on the extent of cambial damage, the disease may be limited to only small portions of the trunk or single branches, and the tree can survive, or the cambium is completely encircled, thereby resulting in death of the tree. Most of the damage by this disease was reported for *A. pseudoplatanus* in urban areas of Great Britain following unusually hot and dry summers around 1950. Therefore predisposition seems to be a major factor, especially since the fungus is present endophytically in healthy, symptomless bark (Gibbs 1998). As a consequence of the draught year 2003 sooty bark disease was found widespread in Germany on several places and different Acer species, at least in the warmer regions (Metzler 2006, Robeck et al. 2008).

Leaf diseases

Generally speaking, most leaf diseases do not belong to those tree disorders of life-threatening potential. However, they can cause reductions in increment and may predispose trees to more serious disorders. Leaf fungi appear to have a special affinity to the genus *Acer*, therefore a few of the more conspicuous leaf disease species will be discussed here in order to enable precise identification. More thorough information is provided by Wulf (1994).

Tar spot disease caused by the ascomycete *Rhytisma acerinum* is certainly one of the most conspicuous leaf diseases of maple. Starting as small yellow infection spots in June, the fungus causes black, sclerotic spots with yellow borders by late summer. On *Acer pseudoplatanus* these are very common and have a diameter of approximately 1 cm. On *A. platanoides* the disease is encountered less often and the spots are up to 2 cm in diameter. During the growing season the conidia of the *Melasmia* anamorph, measuring 5 × 1 µm, are produced on the black spots, whereas the teleomorph is produced on fallen leaves during the saprophytic phase in the following spring. From May onward the filiform ascospores measuring 2 × 60 µm are ejected from the leaf litter and infect the upper epidermis of the newly formed leaves. The presence of tar spot disease was formerly interpreted as a sign of clean air, since it is found less often in urban areas with higher amounts of SO_2 than in forests. However studies have shown that there is no correlation between air pollution and tar spot disease. It is rather the removal of fall leaves in urban settings which serves to reduce the inocolum potential for infection in the following spring (Leith and Fowler 1987).

Also common on maple leaves are parenchyma galls, especially the window galls caused by the window gall midge *Dasineura vitrina*, which can be very common in some years. The galls are round, approximately 1 cm in diameter and produce a central window in the form of a small membrane on the leaf underside which has led to the name of the disorder. The galls are often colonized by leaf fungi which prevent complete development of the gall and cause the aborted tissue to be shed, thereby leading to a shothole appearance of the leaf. The fungi responsible for the elimination of the galls, e.g. *Diplodina acerina* and other mainly endophytic fungi, are seldom a parasite of the tree. Instead they can be interpreted as mutualistic symbionts on account of their active role in defense against pest insects (Wulf 1990).

Asian longhorned beetle (ALB)

In 1996 the death of many *Acer platanoides* trees was registered in New York, USA. The damage was caused by a cerambycid beetle introduced from Asia, *Anoplophora glabripennis* Motschulsky, which was then named Asian Longhorned Beetle (ALB). The large, up to 35 mm long beetle deposits its egg masses directly into small oviposition wounds on the tree bark. The larvae penetrate first the bark and then deeper into the sapwood. The large feeding galleries measure up to 8 mm across and somewhat resemble those of the Leopard moth (*Zeuzera pyrina*). The infested tree parts soon die, accompanied by rapidly developing wood decay.

ALB, which is spread with wooden packing material and infested tree parts, poses a serious threat especially to amenity trees, but possibly also to forest trees, because of its wide host range. Since it prefers *Aesculus* and *Acer*, the latter genus would be most at risk as a forest tree. Therefore, early warnings on the possible introduction were given and quarantine measures were recommended (Wulf 1999). Although these have since been implemented, the introduction of ALB to Europe was not prevented, and ALB infestations were found 2001 in Austria and 2004 in Germany (BBA 2003; Hamberger 2004). Control measures are currently focusing on complete eradication of the pest in those localities where ALB was found, every single tree with symptoms being removed and destroyed. At the same time, the public has been informed on the threat in order to rapidly find possible further infestation sites (Wulf and Pehl, 2002). During the last years different introductions of Anoplophora chinensis, a near relative species to ALB with a broader range of host trees, could not be prevented. In Italy the infested region already covers more than 100 km (Schröder and Maspero 2008).

Other disorders of maple are presented in Tables 3.3 and 3.4.

3.2.3 Common ash

On better sites *Fraxinus excelsior* produces good quality wood. Without doubt it is one of the more valuable broadleaved trees in Central and Western Europe and has also gained importance in urban areas. Its designation as "tree of the year" in 2001 in Germany initiated a more thorough examination of disease-causing organisms of this tree species. (Schröder and Dujesiefken 2001).

Ash canker

Canker diseases of ash branches and trunks belong to the most important and conspicuous disorders of *F. excelsior*. With regard to causal organisms and symptoms, two different canker types must be differentiated: on one hand, a fungal disease caused by *Neonectria galligena* (Bres.) Rossmann and Samuels (Rossmann et al. 1999) and on the other, a bacterial disease caused by *Pseudomonas syringae* subsp. *savastanoi* pv. *fraxini* Janse.

Several fungal species of the genus *Nectria* cause characteristic canker diseases on a variety of tree species. A typical characteristic of *Nectria* cankers is the

Table 3.3. Fungi disorders of maple.

Species	Symptoms	Reference
Verticillium Wilt *(Verticillium dahliae)*	Problem of nurseries and ornamental trees. Wilting of leaves and shoot tips on scattered branches which can later die completely. Greenish brown stains in the sapwood.	Butin 1995
Giant Leaf-blotch *(Pleuroceras pseudoplatani)*	Large brownish blotches (2–5 cm) with blackish vein necrosis on the underside of the leaf.	Butin 1995, Wulf 1988
Cristulariella Leaf Spot *(Cristulariella depraedans)*	Grey-white roundish 0.2–1.0 cm spots in large numbers evenly distributed. Stalked spore heads, 0.1–0.2 mm in diameter, formed on the underside of the leaves.	Butin 1995, Wulf 1994
Powdery Mildew *(Sawadea spp.)*	Fine white mycelium coating the upper leaf surface of *A. platanoides* (*S. tulasnei*) or on both leaf surfaces of maple (*S. bicornis*).	Butin 1995, Wulf 1994

relatively symmetrical form of the resulting trunk defect. The cankers often arise where dead twigs or branches join the main trunk. From there oval bark and cambium lesions develop into open, crater-like trunk cankers which reach up to 30 cm in diameter after several years (Butin 1995). A typical sign of these cankers are the elliptical ribs of annual woundwood produced by the tree, which can serve to determine the canker age and led to the term "target canker". The symmetrical shape is caused by the perennial alternation between attacks of the fungus on the cambium (usually during the tree´s dormancy) and the defense reactions of the tree in the form of callus and woundwood production, during the growing period. This struggle can continue for years without either of the opponents gaining an advantage over the other so that extensive cratered rings are produced.

Neonectria galligena is an ascomycete of worldwide distribution which occurs as a parasite not only on ash but also on other broadleaved forest trees such as maple, birch and poplar. It is of special economic concern on apple trees grown for fruit production. The fungus occurring on ash is termed *N. galligena* f. sp. *fraxini* suggested by Flack and Swinburne (1977) in order to underline the specific nature of those strains. A reliable disease diagnosis is enabled by searching for the fruit bodies of the causal fungus, although the typical shape of the cankers is usually sufficient to reach a correct diagnosis. The pin-sized groups of red perithecia develop at the margin of the dead, last-formed callus ring. They contain two-celled, hyaline ascospores measuring 22 × 6–9 µm (Rossmann et al. 1999, Perrin 1976). The dead wound margins as well as the discoloured xylem are also later colonized by saprophytic fungi (Butin 1995). In recent years, methods for molecular detection of *N. galligena* and differentiation from other *Nectria* species have been developed (Langrell 2002, Plante et al. 2002).

Table 3.4. Insect disorders of maple.

Species	Symptoms	Reference
Sycamore moth (*Acronicta aceris*)	Larvae (35 mm long when fully grown, bearing long tufts of yellow or reddish hairs) feed on leaves of sycamore, horse chestnut and other deciduous trees.	Alford 1991, Schwenke 1978
Leopard moth (*Zeuzera pyrina*)	Larvae yellowish with prominent black pinacula, up to 60 mm long. Enters larger twigs and branches to feed in the heart wood. Galleries extend for 40 cm and more. Break or rot of trees can follow. More important for amenity trees than forest trees.	Alford 1991, Schwenke 1978, Schwerdtfeger 1981
Maple leaf roller weevil (*Deporaus tristis*)	Adult beetle feeds on the leaves, removing significant amounts of tissue and forms characteristic tightly wound leaf rolls within which larval development takes place. Widely distributed in southern Europe.	Alford 1991, Schwenke 1974
Sycamore leaf gall mite, Maple bead-gall mite (*Artacris* spp.)	Elongated horn-formed (*A. macrorhynchus*) or bead-shaped (*A. cephaloneus*) red galls on the upper surface of sycamore or maple leaves.	Alford 1991, Buhr 1964
Eriophyes-Galls *Eriophyes* spp.)	Forming yellowish to brownish felt-like galls on the underside of the leaves.	Alford 1991, Buhr 1964

In contrast to fungi, which cause regular canker wounds starting from the original point of infection, bacterial tree cankers are characterized by a very irregular appearance. The bacterial canker of ash caused by *Pseudomonas syringae* subsp. *savastanoi* pv. *fraxini* is no exception and shows typical bark excrescences which are the result of repeatedly interrupted wound reaction processes (Janse 1981). The bacteria gain entrance into the bark parenchyma not only through wounds but also through leaf abscission points or lenticels. The damage to cambial cells then results in swelling and cracking of bark, thereby initiating further wound reactions. The multiplication of the bacteria, associated with slime production, further interferes with the wound reactions and leads to the typical, irregular canker symptoms (Riggenbach 1956). A multitude of clustered, black, necrotic bark regions is the result. At first glance, and in their early stages, they can be confused with feeding symptoms of adult ash bark beetles (*Leperesinus varius,* see Table 3.6). Advanced stages of the disease are characterized by colonization of the necrotic regions by fungi, especially those of the genera *Fusarium* und *Phoma*, which can compound damage to the tree (Butin 1995).

Both canker-causing organisms of ash usually do not result in lethal damage to trees, but rather cause an extensive loss of wood value. From a forestry point of view it is therefore necessary to undertake control and prevention measures. In this context it is important to avoid growing ash on suboptimal sites since bacterial canker in particular arises on acidic, boggy and extremely wet soils. Early cutting of affected trees is also recommended in order to prevent further spread of the disease (Janse 1981). Resistance trials have also been carried out, and with respect to the bacterial canker, the planting of more highly resistant ash clones seems promising (Dam and Voet 1991).

Ash dieback

Since a few years damage on different ash species (*Fraxinus excelsior, F. angustifolia*) has been increasingly recorded from Middle and Northern Europe. Retarded sprouting and crown dieback of young and old trees in forests and nurseries were reported. The pathogenesis is chronic but often lethal after some time of suffering. The causes of the decline are very different or complex and can be attributed to abiotic predisposition and biotic factors. Among them a vascular fungus (*Chalara fraxinea* Kowalski), recently described in Poland and in the meantime also recorded in Germany, may have significance (Kowalski 2006, Schumacher et al. 2008).

Ash bud moth

The ash bud moth, *Prays fraxinella,* is a monophagous insect of *Fraxinus excelsior* which occurs in all of Europe except the extreme Southeast. In Northern Europe it produces only one generation annually, whereas it produces two generations per year in the other regions. The main damage is caused by those individuals of the second generation that survive the winter. Eggs are deposited on the leaves and the young larvae feed in leaf mines, whereas older stages destroy all leaf layers except the lower epidermis. Feeding by old larvae leads to typical feeding holes in leaves which are joined together by webs. Pupation occurs in loose cocoons in the upper crown or between fallen leaves on the forest floor.

The young larvae of the second generation leave the mines before leaf fall and infest the terminal buds of branches to over-winter. In early spring they destroy the inner bud tissue before emerging to feed on newly formed leaves (Schwenke 1978). Overwintering of larvae in bark fissures and infestation of buds in freshly formed terminal shoots in spring has also been reported (Alford 1991). In either case the destruction of the terminal shoot results in twin, or codominant, stems due to the equal growth of the remaining latteral buds. The production of twin leaders due to moth damage is more common in ash than on any other broadleaved tree and poses considerable damage to forest production of valuable timber. On older trees control measures are hardly practical and are also unnecessary. In young trees insecticide treatments against larvae of the first generation are possible in June and against larvae of the second generation in September. In young stands and plantations the removal of one of the twin stems can be recommended following years with extensive damage.

Other disorders of ash are presented in Tables 3.5 and 3.6.

Table 3.5. Fungi disorders of ash.

Species	Symptoms	Reference
Verticillium Wilt (*Verticillium dahliae*)	Wilt, premature leaf loss and crown dieback. Mainly on ornamental trees.	Hiemstra 1995
Ash powdery mildew (*Phyllactinia fraxini*)	White mycelial coating on the leaf underside.	Nienhaus and Kiewnick 1998
Inonotus hispidus	Basidiomycete with reddish-brown consoles 20–30 cm in diameter in different heights of the trunk or on large branches. Annual fruit body drying to a blackish color. Invades trees through wounds and causes white rot. Also common on Plane-tree (*Platanus* spp.).	Butin 1995
Perenniporia fraxinea	Basidiomycete originating in the root system and causing a white rot of large roots and lower parts of the trunk. Upper side of fruit bodies brownish to greenish (algae), pore surface cream white. Fruit bodies develop mainly between the large roots at the base of the tree. Also found on *Robinia* and *Ulmus*.	Kehr et al. 2000

3.2.4 Wild cherry

In European forests *Prunus avium* is among the most valuable veneer species if trees are allowed to realize their great growth potential in younger years (up to approximately 60 years). As wild cherry is a pioneer species dependent on good light conditions, neighboring trees must be thinned regularly to ensure satisfactory crown formation. Branch pruning may also be carried out to ensure a clean bole. Trees lacking good crown development will not produce a valuable bole and are also more susceptible to secondary parasites and abiotic disorders. Along with birch, alder and poplar, wild cherry is one of those forest species with only a very limited life span. Therefore it becomes increasingly more susceptible to secondary pathogens and root and butt rots at an older age. In older stands and trees with large branch wounds, the fungus *Laetiporus sulphureus* is a common cause of decay. Wild cherry is to a certain degree susceptible to many of the diseases affecting *Prunus* species in fruit production (Ogawa et al. 1995).

Gummosis

Gummosis is a disease condition in which affected trees produce an excess of gum from buds, branches and trunks. Many causes such as wounding, extremely fast growth due to an excess of fertilizer and water and also some bacterial and fungal

Table 3.6. Insect disorders of ash.

Species	Symptoms	Reference
Lilac leaf miner (*Caloptilia syringella*)	Young larvae mine in leaves. The galleries soon widen into extensive blotches where several larvae usually feed together. Later the larvae feed within distinctive frass-filled shelters formed from webbed down leaves.	Alford 1991, Schwenke 1978
Ash gall midge (*Dasineura fraxinea*)	Round parenchyma-gall on lower leaf surface. Affected tissue can fall out leaving shot holes. Heavy infestation can cause death of young trees.	Buhr 1964, Schwenke 1982
Ash gall mite (*Aceria fraxinivora*)	Conspicuous galls on flowers and young fruit tissues which remain in the tree after leaf fall.	Buhr 1964, Schröder 2002
Leopard moth (*Zeuzera pyrina*)	See maple.	Alford 1991, Schwenke 1978
Ash bark beetle (*Leperesinus varius*)	Bark breeding beetle with typical double-armed crossbreeding gallery. On felled trees and pre-damaged older trees. Maturation feeding of beetles on younger trees causes symptoms similar to beginning infection of bacterial canker.	Altenkirch et al. 2002, Schwenke 1974
Hornet (*Vespa crabro*)	Feeds on the bark and cambial layer of young twigs and branches predominantly in autumn in order to utilize the bark exsudates. Production of wound periderm. Death of leaders when whole circumference is stripped of bark, promoting wind damage and formation of twin leaders.	Brauns 1991, Bunn 1988, Santamour et al. 1986, Schwenke 1982

diseases can lead to gumming which is a defense reaction of the tree (Pscheidt and Byther 2001, Hartmann et al. 1995). Bacterial cankers caused by *Pseudomonas syringae pv. avii* (Synonym: *Pseudomonas mors-prunorum*) can cause severe gummosis. In order to prevent this condition it is important to reduce the mechanical and pathological causes of branch and trunk injury to a minimum and promote moderate rather than excessive growth.

Crown gall

Crown gall is a disorder of the roots and sometimes parts of the crown caused by the bacterium *Agrobacterium tumefaciens* which parasitizes many different plant species. On young branch tissues the galls are first soft. On older tissues they become hard and

Table 3.7. Other disorders of wild cherry.

Species	Symptoms	Reference
Monilia-disease *(Sclerotinia laxa)*	Fungal disease which kills flowers, leaves and young shoots from the tip downward. Pathogen of many species and clones of Prunus.	Nienhaus and Kiewnick 1998, Nienhaus et al. 1996
Shot hole disease *(Stigmina carpophila)*	Fungus causing many 1–2 mm large leaf spots, at first reddish, then dark brown. Conidia of the Clasterosporium anamorph formed on necrotic tissue. The dead tissue later falls out, leading to a shot hole appearance.	Butin 1996, Hartmann et al. 1995
Leaf anthracnose (*Apiognomonia erythrostoma*)	Fungus disease causing large leaf blotches, first yellowish, then brown. Dead leaves are not shed in the same year.	Butin 1996
Spray-disease of leaves *Blumeriella jaapii*	Small reddish-violet spots on leaves, leading to large necrotic areas later in the year. Conidia of the *Phloesporella* anamorph in acervuli on underside of leaves.	Butin 1996, Hartmann et al. 1995
Armillaria root disease *Armillaria mellea* s.l.	This fungus kills roots and cambium and also causes white rot decay of roots and wood. Symptoms consist of gummosis at the base of the tree, white mycelial fungal fans in the cambial region and wilting and dieback of the crown at later stages. *Armillaria* fruit bodies may be found at the base of the tree in autumn. Problem especially acute in plantations.	Fox 2000, Shaw and Kile 1991

fissured and can measure up to several cm across. Intense gall formation can lead to the death of young trees if the galls completely encircle the tree. *A. tumefaciens* is mainly a problem in nurseries and young plantation trees (Pscheidt and Byther 2001). It is present in soil and can spread from infected nursery sites. It can be prevented in part by reducing wounding since the bacteria enter wounds caused by cultivation and planting procedures. If wild cherry is planted from nursery stock, it is important to ensure that the stock does not carry the disease.

3.2.5 Walnut

The genus *Juglans* is represented in Europe by several species and the natural hybrid *Juglans* x *intermedia*, a hybrid between *J. nigra* und *J. regia*. From an economic perspective *J. regia, J. nigra* and *J.* x *intermedia* are important forest and plantation species. *J. regia* is an important fruit tree and also produces valuable wood. *J. nigra* was brought to Europe from North America several hundred years ago and is a prized forest and amenity species. Walnut wood is the most valuable of the forest trees growing in Europe (Kucera 1991) and is potentially especially valuable in the future because the dark-colored wood may be able replace some dark tropical woods. Wood quality in older trees can be negatively influenced by the decay fungus *Inonotus hispidus*, which is also known from *Fraxinus, Platanus, Sorbus* etc.

Juglans is damaged by both abiotic and biotic factors such as late spring frosts on some sites and deer browsing and mouse and rabbit injury to young stems. In Europe *Nectria* canker disease is a problem in some plantations and forest stands. Butternut canker, which is not yet present in Europe, could potentially wipe out walnut in many areas. Therefore these two diseases will be discussed here in more detail.

Nectria canker

In some species and hybrids of walnut, especially *Juglans x intermedia*, the ascomycete *Nectria coccinea* causes a serious canker disease of young trees which can lead to high mortality rates within the first ten years after planting (Schröder et al. 2002). The cankers are similar to those described for *Neonectria galligena* on Ash, i.e. a target canker type characterized by perennial ribs of woundwood, leading to large, open cankers. Often a branch stub in the center of the canker indicates where the wound pathogen invaded the main stem. In wet spring weather the red perithecia of *N. coccinea* with two-celled spores measuring 12–15 × 5–6 µm can be seen on the last-killed ribs of woundwood. In addition, fruit bodies of the *Cylindrocarpon* anamorph may be present. In weakened trees cankers can quickly girdle the main stem thereby causing death of the whole tree or of the leader. If the leader is killed, sprouting and codominant stems may appear which greatly reduces wood quality. Apparently disease incidence is encouraged by high supplies of nitrogen and on sites with late spring frosts and extremely high growth rates which lead to bark fissures and thus entry points. Wounding of trees, especially by mice and deer but also by intense pruning measures, should be kept to a mininmum since the fungus is a known wound pathogen.

Sirococcus canker

Since its introduction to North America approximately 45 years ago, the fungus causing Butternut canker, *Sirococcus clavigignenti-juglandacearum* Nair, Kostichka & Kuntz, has led to the extensive destruction of Butternut (*Juglans cinerea* L.) in its native host range. The fungus causes perennial branch and trunk cankers which can completely girdle the cambium and kill the host. The pathogen is not yet present in Europe, but the main walnut species cultivated here, *Juglans regia* L. and *J. nigra* L., are also susceptible and would be seriously endangered both in fruit plantations and in forests as well as in urban areas, should the fungus be introduced (Schröder et al. 2002, Ostry and Katovich 1997, Orchard et al. 1982). Symptoms consist of branch and crown dieback of Butternut trees of all ages on different types of sites. The perennial cankers occur on branches as well as on the main trunk. Young cankers are elliptical, often with a black center and a white margin. The fungus produces a thin, black stroma under the dead bark on which 1 to 4 mm long hyphal pegs are produced. The pycnidia produce conidia measuring 9–17 × 1–1.5 µm. Leaf scars, buds and twigs serve as entry points for the wound pathogen which does not possess a known teleomorph (Furnier et al. 1999).

Sirococcus clavigignenti-juglandacearum is transmitted by several insect species, wind, rain and seeds which is very relevant to nurseries (Innes 1997). During the stratification of seeds latent infections of the fungus can also survive in cotyledons for longer periods, thus making it essential to take precautions when importing walnut seeds from areas of the U.S. where the disease is present. Of all known walnut diseases, *Sirococcus* canker certainly poses the greatest potential threat to walnut cultivation for forest and fruit production in Europe.

3.2.6 Chestnut

European Chestnut, *Castanea sativa*, is a species which originated in the eastern Mediterranean and Black Sea Regions but has been present in southern and Western European countries for thousands of years. Even north of the Alps, its history of cultivation extends back approximately 2000 years and plays an important part as a landscape and forest element in some regions, for instance in the wine-growing climate of Southern Germany. Chestnut is a multi-use tree species which is still very important to rural areas especially in Southern Europe. It is valued for its diverse products, from edible seeds to furniture made from the heartwood. The future potential of Chestnut as a species suited for modern agroforestry has been recognized, and the The European Chestnut Network aims to work on improving genetic quality and promote the planting of chestnut. Two devastating diseases of *Castanea sativa* which could greatly limit the success of its cultivation will be discussed here.

Chestnut blight

The causal fungus of Chestnut blight, *Cryphonectria parasitica* (Murr.) Barr (Anamorph *Endothiella parasitica* [Murr.] Roane), originates in Eastern Asia where it is a weak parasite of *Castanea mollissima* and *C. crenata.* It was introduced to North America in 1904 and in the process of several decades destroyed most of the extremely susceptible American Chestnut (*Castanea dentata*) in eastern North American forests (Griffin and Elkins 1986). In Europe it was observed for the first time in 1938 and has since spread northward, crossing the Alps in the 1980s (Bisegger and Heiniger 1991). Since around 1985 the fungus has also been present at several locations in Southern Germany (Seemann and Zajonc 1994) and is apparently progressing northward (Seemann et al. 2001, Kehr 1997, Heiniger and Rigling 1994, Seemann and Unger 1993). The fungus is a wound parasite and is disseminated by wind, rain splash, birds and various insect species which inhabit chestnut bark.

Symptoms consist of wilting and flagging of leaves and branches in summer which is caused both by toxins of the fungus and death of cambium due to canker activity. Fresh bark cankers are reddish-brown and have a watery growth zone. The fungus produces characteristic cream-colored mycelial fans in the cambial region. Tree parts above completely girdling cankers die and vigorous sprouting activity may occur beneath the canker zone. Older cankers representing chronic disease forms on thicker branches and the trunk possess a cracked and swollen appearance. The *Endothiella* anamorph of the fungus is produced in orange pycnidial pustules on freshly killed bark in late summer and winter and the hyaline, elongate conidia measure 3.5×1.2 µm. The teleomorph is found irregularly on dead parts the following year and produces two-celled ascospores measuring $7–11 \times 3.5–5$µm.

C. parasitica is characterized by the presence of groups with different vegetative compatibility (v-c groups). Within the groups a phenomenon known as hypovirulence, which is in effect an infection of the fungus by virus-like particles, is transmissable by mycelial contact (Choi and Nuss 1992, Day et al. 1977). Hypovirulent strains of the fungus are less aggressive and chestnut trees are able to survive the attack and form effective wound barriers around chronically cankered bark regions thereby leading to noticeable swelling and cracking, but not death, of the affected parts (Heiniger and Rigling 1994). Hypovirulence is transmissable only within certain v-c groups, which could somewhat limit the success of artificial introduction of hypovirulent strains into diseased stands. However, recent research indicates that the early introduction of hypovirulence can be recommended (Hoegger et al. 2003).

Chestnut blight is a good example of a devastating invasive disease resulting from contact of a susceptible host with an aggressive new pathogen from outside the natural range of the host tree. For *Castanea dentata* resistance breeding appears to be the only hope of bringing the tree back into the forests. European chestnut, *C. sativa*, is not quite as susceptible as *C. dentata* and hypovirulence was able to spread more quickly in European stands, therefore the disease impact in Europe was not as serious as in North America. Nevertheless, Chestnut blight still poses a significant threat to the production of valuable chestnut wood in the future since the occurrence of bark lesions on the main stem reduces wood quality. The future of chestnut cultivation in Europe, including cultivation as a fruit tree, will certainly depend on the degree of success with which natural hypovirulence of the fungus is able to spread, thereby reducing the threat posed by the disease.

Ink disease of European chestnut

This disease is similar to *Phytophthora*-disease of alder and is caused by *Phytophthora cambivora* and in some cases other *Phytophthora* species, especially *Ph. cinnamomi*. It has been causing widespread damage to *Castanea sativa* in several European countries such as Italy, France and especially Spain possibly for several hundred years (Peace 1962) and in some regions is certainly an even bigger problem than Chestnut blight. The fungus infects the plants through the roots and gradually moves upward into the lower parts of the trunk, killing the cambium and causing typical dark, tongue-like discolorations of the bark at the base of the tree accompanied by slime flux (Grente 1961). The dark discolorations have led to the common name for the disease. The crowns of infected trees eventually exhibit dieback symptoms and wilting, resulting in the death of the tree. Site conditions, especially soils with fluctuating water levels, apparently contribute to disease incidence. Asian chestnut species such as *C. crenata* are fairly resistant. The disease can be transferred with infected soil therefore special care must be taken to use healthy planting stock from uninfected nurseries when planting chestnut for wood or fruit production.

3.2.7 Important disorders of other valuable broadleaved trees

Dutch elm disease

Dutch elm disease is certainly the most important wilt disease of trees to date and has led to the death of many millions of elms in forests and urban areas of North America and Europe (Stipes and Campana 1981). American elm (*Ulmus americana*) and English elm (*U. procera*) are extremely susceptible to the disease. The continental European species such as *U. glabra, U. carpinifolia* and *U. laevis* are also highly to fairly susceptible, whereas Asian elm species are generally more resistant (Brasier 1991). The disease effectively prevents the planting of non-resistant elm species and clones and endangers the remaining elms found in European forests. This is especially unfortunate since elms such as *Ulmus glabra* produce straight, clean boles and much sought-after dark veneer wood which in the long term could be able to replace some of the dark tropical hardwoods that are currently being harvested at an alarming rate. This wood characteristic makes elm (and walnut) among the potentially most valuable broadleaved tree species in European forests.

Dutch elm disease is caused by several species and races of fungi belonging to the *Ophiostoma ulmi* complex. The aggressive strain responsible for the recent damage to elms in Europe, Asia and North America is *Ophiostoma novo-ulmi* Brasier. Another less aggressive strain, *Ophiostoma ulmi* (Buism.) Nannf., is also found throughout the natural range of elms. In recent years other strains have been identified which may have played a role in the development of the aggressive strains in North America and Europe (Brasier and Buck 2002, Hoegger et al. 1996, Brasier and Mehrotrova 1995, Brasier 1991). The history of Dutch elm disease is well documented (Butin 1995, Phillips and Burdekin 1982). The disease apparently originates in Asia and was first observed in Europe in the early 1900s. It was introduced to North America some time around 1929. After causing a first destructive wave of disease, some progress was made with resistance screening and resistance breeding. The new "resistant" clones as well as most of the remaining natural and urban elm stands were then destroyed by a second pandemic caused by the above mentioned more virulent forms of the fungus.

Disease symptoms consist of sudden wilting and yellow discoloration of leaves of individual twigs and branches in the tree crown. The symptoms can also progress downward in the tree, killing it within one to several years, but in the case of infection by *O. ulmi*, chronic disease forms are also known to occur. The infection by the fungus occurs primarily through wounds and through root anastomoses between neighboring trees (Sinclair et al. 1993). The fungus grows actively in the xylem vessels of the tree and is also passively disseminated in a yeast-like form by xylem sap flow. Toxins are produced which cause physiological damage to the vessels and promote tylosis production, i.e. occlusion of vessels by the host tree, leading to water stress and wilting of affected branches. After death of the branch the anamorph of the fungus, *Pesotum ulmi* (Schwartz) Crane & Schoknecht, is produced on dead tissues and in bark beetle breeding galleries. The teleomorph is produced less often on dead bark and wood.

The most important vectors of the disease are bark beetles of the genus *Scolytus* which breed in dead and dying trees. The young beetles are contaminated with spores and spread the fungus to uninfected trees when they carry out their maturation feeding, mainly in the branch crotches of small twigs and branches. The presence of dying and dead trees in which the bark beetles may breed is the single most important factor affecting the spread of the disease. Therefore one of the most effective control measures against Dutch elm disease is to continually remove such breeding material, especially in the winter months, so that young fungus-infected beetles cannot hatch in spring. Other control methods such as the injection of systemic fungicides can succeed in keeping valuable urban trees disease-free for several years, although this is not a feasible method in forest and plantation trees.

In the future resistance breeding of elms for good growth characteristics and moderate to high disease resistance will undoubtedly progress. However, in light of the ability of the fungus to rapidly adapt and form new strains, it is difficult to forecast if the resistant elm clones can remain disease-free for the long periods necessary for forest production cycles. In urban areas, the disease can be contained to some extent by implementing strict hygienic management (Mackenthun 2004, Tainter and Baker 1996).

Phytophthora disease of alder

Since the 1980s there have been increasing reports of tree diseases caused by the fungi-like organisms of the genus *Phytophthora*. These primary, soil-inhabiting pathogens usually cause damage to living roots and cambium in the lower trunk regions and in some cases also leaf and twig dieback, as is the case with Sudden Oak Death (Werres 2001, 2000, Werres et al. 2001, Werres and Marwitz 1997). Depending on the fungal species and its specificity, only a single host plant species or many different host plant species can be affected (Werres and Hahn 2000). *Phytophthora* disease of alder is a good example of these diseases and, along with *Phytophthora cambivora* disease of *Castanea sativa*, it has probably had the greatest impact of this disease category in Europe (Gibbs et al. 1999). First reports of damage on *Alnus glutinosa* were registered in Great Britain (Brasier et al. 1995, Gibbs 1994) and in the following years the disease was found also in Germany (Hartmann 1995) and Austria (Cech 1997), among other European countries. *A. glutinosa* is the primary host, although *A. incana* and *A. cordata* have also been reported as hosts. Isolations from diseased trees yield known *Phytophthora* species such as *P. megasperma* and *P. citricola* in addition to a new species *Phytophthora alni* which could be a hybrid of previously known species (Werres 2000, Brasier et al. 1999).

The most conspicuous symptom of *Phytophthora* is death of tree crowns in summer. This is preceded by the production of smaller leaves which are lacking in chlorophyll and thus are of a lighter yellowish color. Since such symptoms can also have other causes, it is important to diagnose the characteristic bark and cambium damage at the base of the tree. The damaged bark regions exhibit soft, wet bark with black spots and slime flux. Removal of the bark reveals the typical longitudinal necroses which arise at the base of the tree and can reach upwards to a height of several meters. They are clearly delimited from healthy bark regions. Young trees often die within a short period of time after attack, but older trees can survive for years and produce extensive sprouting from lower parts of the trunk.

Pathogenesis of the disease is not completely understood but it seems very probable that the infection process begins in the soil and that roots are first infected before the fungus is able to advance upwards. During this process the cambium is killed and the younger part of the xylem is invaded. The destruction and blockage of the water vessels increasingly submits the tree to water stress so that prolonged periods of drought quickly lead to the death of heavily infected trees. On the other hand, sufficient water supply often leads to chronic disease progression.

Water is a key factor in the disease also because *Phytophthora* species form sporangia and mobile zoospores in the presence of high moisture, especially in the presence of flowing water. The spread of the disease is therefore promoted by high soil moisture, stagnating water levels, periodic flooding and proximity to flowing water. Direct control measures using fungicides are not a realistic alternative. Therefore, for the time being, planting of alder on disease prone sites, especially those with low pH-values, should be viewed critically. As a control measure for infested stands it is advised to cut affected trees and thus promote the growth of healthy sprouts at the base of the tree. The sprouts must, however, be checked for further spread of the disease on a regular basis. The role of resistance in various alder clones has yet to be determined as a possible basis for keeping the disease incidence under control. Since alder is one of the few tree species suited for planting on many wet sites, resistance breeding offers some hope for the future. In heavily damaged stands individual trees survive the disease and could thus be a basis for resistance breeding.

References

Alford, D.V. 1991. Pests of ornamental trees, shrubs and flowers. Wolfe Publishing Ltd. London.

Altenkirch, W., Majunke, C. and Ohnesorge B. 2002. Waldschutz auf ökologischer Grundlage. Ulmer, Stuttgart.

BBA 2003. Symposium in der BBA: Der Asiatische Laubholzbockkäfer – ein Risiko für den Baumbestand. Nachrichtblatt des Deutschen Pflanzenschutzdienstes 55(4): 65–98.

Bisegger, M. and Heiniger, U. 1991. Chestnut blight (*Cryphonectria parasitica*) north of the Swiss alps. European journal of forest pathology 21: 250–252.

Brandenburger, W. 1985. Parasitische Pilze an Gefäßpflanzen in Europa. Gustav Fischer Verlag, Stuttgart.

Brasier, C.M. 1991. *Ophiostoma novo-ulmi* sp. nov., causative agent of current Dutch elm disease pandemics. Mycopathologia 115: 151–161.

Brasier, C.M. and Buck, K.W. 2002. Rapid evolutionary changes in a globally invading fungal pathogen (Dutch elm disease). Biological Invasions 3: 223–233.

Brasier, C.M., Cooke, D.E.L. and Duncan, J.M. 1999: Origin of a new Phytophthora pathogen through interspecific hybridization. Proc. Natl. Acad. Sci. USA 96: 5878–5883.

Brasier, C.M. and Mehrotrova, M.D. 1995. *Ophiostoma himal-ulmi* sp. nov., a new species of Dutch elm disease fungus endemic to the Himalayas. Mycological Research 99: 205–215.

Brasier, C.M., Rose, J. and Gibbs, J.N. 1995. An unusual Phytophthora associated with widespread alder mortality in Britain. Plant Pathology 44: 999–1007.

Brauns, A., 1991. Taschenbuch der Waldinsekten. 4th edition, Fischer Verlag, Stuttgart.

Breitenbach, J. and Kränzlin, F. 1986. Pilze der Schweiz, Vol. 2, Nichtblätterpilze. Verlag Mycologia, Luzern.

Bressem, U. 2002. Nectria-Erkrankung an Bergahorn im Vogelsberg. Allgemeine Forstzeitschrift – Der Wald 7: 368–370.

Browne, F.G. 1968. Pests and diseases of forest plantation trees. Clarendon Press, Oxford.

Buhr, H. 1964. Bestimmungstabelle der Gallen an Pflanzen Mittel- und Nordeuropas. Band 1 VEB Gustav Fischer Verlag, Jena.

Bunn, D.S. 1988. Observations on the foraging habits of hornet *Vespa crabro* L. (Hym., Vespidae). Entomologists monthly magazine 124: 187–194.

Butin, H. 1995. Tree diseases and disorders. Oxford University Press, New York, Tokyo.

Butin, H. 1996. Krankheiten der Wald- und Parkbäume. Thieme, Stuttgart, New York, 3. edition.

Cech, T.L. 1997. Phytophthora-Krankheit der Erle in Österreich. Forstschutz Aktuell 19/20: 14–16.

Choi, G. H. and Nuss, D. L. 1992. Hypovirulence of chestnut blight fungus conferred by an infectious viral cDNA. Science 257: 800–803.

Dam, van, B.C. and Voet, van der, H. 1991. Testing *Fraxinus excelsior*, *Fraxinus americana* and *Fraxinus pensylvanica* for resistance to *Pseudomonas syringae* subsp. *favastanoi* pv. *fraxini*. European journal of forest pathology 21: 365–376.

Day, P. R., Dodds, J. A., Elliston, J. E., Jaynes, R. A. and Anagnostakis, S. L. 1977. Double stranded RNA in *Endothia parasitica*. Phytopathology 67: 1393–1396.

Ellis, M. B. and Ellis, J. Pamela. 1985. Microfungi on Land Plants. An Identifaction Handbook. Croome Helm, London.

Flack, N.J. and Swinburne, T.R. 1977. Host range of *Nectria galligena* Bres. and the pathogenicity of some Northern Ireland isolates. Transactions of the British Mycological Society 87: 185–192.

Fox, R.T.V. (ed.) 2000. Armillaria root rot: Biology and control of Honey fungus. Intercept/ Andover.

Furnier, G.R., Stolz, A.M., Mustaphi, R.M. and Ostry, M.E. 1999. Genetic evidence that butternut canker was recently introduced into North America. Canadian journal of botany 77: 783–785.

Gibbs, J.N. 1994. Phytophthora root disease of Common alder. Research Information Note 258, Forestry Authority, Research Division, Alice Holt Lodge, Wrecclesham, Farnham, Surrey.

Gibbs, J.N. 1998. Understanding and controlling diseases od woody plants: an historical perspective. Paper No. 3.7.1S International Congress of Plant Pathology (ICPP) 1998, 9–16. August 1998, Edinburgh.

Gibbs, J.N., Lippscombe, M.A. and Peace, A.J. 1999. The impact of *Phytophthora* disease on riparian populations of common alder. European journal of forest pathology 29: 39–50.

Grente, J. 1961. La maladie de l'encre du chataignier. II. Les agents pathogènes: *Phytophthora cambivora* et *P. cinnamomi*. Annales des épiphyties 12: 25–59.

Griffin, G.J. and Elkins, J.R. 1986. Chestnut blight. In: Roane, m. et al. (eds.) Chestnut blight, other Endothia diseases and the Genus Endothia. APS Mongraph Series, St.Paul. Pp. 1–26.

Hamberger, J. 2004. Bedenklicher Bock aus Braunau bedroht bayerische Bäume. LWF aktuell 45: 19–20.

Hartmann, G. 1995. Wurzelhalsfäule der Schwarzerle (*Alnus glutinosa*) – eine bisher unbekannte Pilzkrankheit durch Phytophthora cambivora. Forst und Holz 18: 555–557.

Hartmann, G., Nienhaus, F. and Butin, H. 1995. Farbatlas Waldschäden. 2. ed. Ulmer, Stuttgart.

Heiniger, U. and Rigling, D. 1994. Biological control of Chestnut blight in Europe. Annual review of phytopathology 32: 581–599.

Hiemstra, J.A. 1995. Verticillium Wilt on Fraxinus excelsior. PhD Thesis, Wageningen Agricultural University.

Hoegger, P.J., Binz, T. and Heiniger, U. 1996. Detection of genetic variation between *Ophiostoma ulmi* and the NAN and EAN races of *O. novo-ulmi* in Switzerland using RAPD markers. European journal of forest pathology 26: 57–68.

Hoegger, P.J., Heiniger, U., Holdenrieder, O. and Rigling, D. 2003: Differential transfer and dissemination of hypovirus and nuclear and mitochondrial genomes of a hypovirus-infected Cryphonectria parasitica strain after introduction into a natural population. Applied and environmental microbiology 69: 3767–3771.

Innes, L. 1997. *Sirococcus clavigignenti-juglandacearum* on Butternut and Black Walnut fruit. Proceedings of the IUFRO WP 7.02.02. Meeting Quebec City, May 25–31, 1997, Foliage, shoot and stem diseases of trees, Canadian Forest Service, LAU-X-122. Pp. 129–132.

Jahn, H. 1990. Pilze an Bäumen. 2. ed., Patzer-Verlag, Berlin – Hannover.

Janse, J.D. 1981. The bacterial disease of ash (*Fraxinus excelsior*), caused by *Pseudomonas syringae subsp. savastanoi pv. fraxini*. I. History, occurrence and symptoms. European journal of forest pathology 11: 306–315.

Kehr, R. 1997. Der Kastanienrindenkrebs – Vorkommen und Bedeutung. Jahrbuch der Baumpflege. Thalacker, Braunschweig. Pp. 110–119.

Kehr, R., Wohlers, A., Dujesiefken, D. and Wulf, A. 2000. Der Eschenbaumschwamm an Robinien – Diagnosemerkmale und Kultureigenschaften. Jahrbuch der Baumpflege 2000, Thalacker Medien Braunschweig. Pp. 200–207.

Kowalski, T. 2006. *Chalara fraxinea* sp. Nov. associated with dieback of ash (*Fraxinus excelsior*) in Poland. For. Path. 36: 264–270.

Kucera, L. 1991. Eigenschaften und Verwendung des Holzes der Walnuß. Allgemeine Forstzeitschrift-Der Wald 12: 608-610.

Langrell, S. R. H. 2002: Molecular detection of Neonectria galligena (syn. Nectria galligena). Myc. Res. 106: 280–292.

Leith, I.D. and Fowler, D. 1987. Urban distribution of Rhytisma acerinum (Pers.) Fries (tar spot) on sycamore. New phytologist 108: 175–181.

Mackenthun, G. 2004. Konzepte zum Erhalt von Ulmen. Jahrbuch der Baumpflege 2004, Thalacker Medien Braunschweig. Pp. 72–82.

Metzler, B. 2006. *Cryptostroma corticale* an Bergahorn nach dem Trockenjahr 2003. Mitt. Biol. Bundestanst. Land-Forstwirtsch. 400: 161–162.

Nienhaus, F., Butin, H. and Böhmer, B. 1996. Farbatlas Gehölzkrankheiten. 2. ed., Ulmer, Stuttgart.

Nienhaus, F. and Kiewnick L. 1998. Pflanzenschutz bei Ziergehölzen. Ulmer, Stuttgart.

Ogawa, J.M., Zehr, E.I., Bird, G.W., Ritchie, D.F., Uriu, K. and Uyemoto, J.K. 1995. Compendium of stone fruit diseases. American Phytopathological Society Press, St. Paul., Minn.

Orchard, L.P., Kuntz, J.E., Kessler, K.J.Jr. 1982. Reactions of Juglans species to Butternut Canker and implications for disease resistance: In. USDA (ed.) Black Walnut for the future. General technical report NC-74. St. Paul, MN. USDA Forest Service, NCFES. Pp. 27–31.

Ostry, M.E. and Katovich, S. 1997. First report of Sirococcus clavigignenti-juglandacearum on Black Walnut. Plant Disease 81(7): 830.

Peace, T. R. 1962. Pathology of Trees and Shrubs. Clarendon Press, Oxford.

Perrin, R. 1976. Clef de détermination des Nectria d'Europe. Bulletin de la Société mycologique de France 92: 335–347.

Phillips, D.H. and Burdekin, D.A. 1982. Diseases of forest and ornamental trees. MacMillan Press, London.

Plante, F.; Hamelin, R.C. and Bernier, L. 2002. A comparative study of genetic diversity of populations of *Nectria galligena* and *N. coccinea var. faginata* in North America. Mycol. Res. 106: 183–193.

Pscheidt, J.W. and Byther, R.S. 2001. *Prunus* Diseases: In: Jones, R.K. and Benson, D.M. (eds.): Diseases of woody ornamentals and trees in nurseries. APS Press, St. Paul, Minn.. Pp. 317–325.

Riecken, I. 1985. Experiments on the prevention of the coral spot disease. Journal of Plant Disease and Protection 92: 516–529.

Riggenbach, A. 1956. Untersuchungen über den Eschenkrebs. Phytopathologische Zeitschrift 27: 1–40.

Robeck, P., Heinrich, R., Schumacher, J., Feind, R. and Kehr, R. 2008. Status der Rußrindenkrankheit des Ahorns in Deutschland. Jahrbuch der Baumpflege. Thalacker Medien, Braunschweig. Pp. 238–245.

Rossmann, A.Y., Samuels, G.J., Rogerson, C.T. and Lowen, R., 1999. Genera of *Bionectriaceae*, *Hypocreaceae* and *Nectriaceae* (*Hypocreales*, *Ascomycetes*). Centralbureau voor Schimmelkultures, Baarn/Delft.

Santamour, F.S., Greene, J.R. and Greene, A. 1986. European hornet damage to ash and birch trees. Journal of Argiculture 12: 273–279.

Schröder, T. 2002. Die Eschengallmilbe *Aceria fraxinivora*. Allgemeine Forstzeitschrift/Der Wald 57: 976–977.

Schröder, T. and Dujesiefken D. 2001. Krankheiten und Schäden der Esche. Allgemeine Forstzeitschrift/Der Wald 6: 276–279.

Schröder, T., Kehr, R. and Hüttermann, A. 2002. Sirococcus-Krebs der Walnuss – eine Gefahr für den europäischen Walnussanbau? Nachrichtenblatt des Deutschen Pflanzenschutzdienstes 54: 5–12.

Schröder, T. and Maspero, M. 2008. *Anoplophora chinensis,* ein naher Verwandter des Asiatischen Laubholzbockkäfers *A. glabripennis* in der Europäischen Union. Thalacker Medien, Braunschweig. Pp. 257–263.

Schütt P., Schuck, H.J. and Stimm, B. 1992. Lexikon der Forstbotanik. Ecomed, Landsberg/Lech.

Schumacher, J., Leonhard, S., Wulf, A. and Heydeck, P. 2008. Neuartiges Eschentriebsterben in Mittel- und Nordeuropa welche Bedeutung kommt dem Gefäßpilz *Chalara fraxinea* sp. nov. zu? Thalacker Medien, Braunschweig. Pp. 145–152.

Schwarze, F.W.M.R., Engels, J. and Mattheck, C. 1999. Holzzersetzende Pilze in Bäumen. Rombach Verlag Freiburg.

Schwenke, W. 1974. Die Forstschädlinge Europas. Band 2, Käfer, Verlag Paul Parey, Hamburg und Berlin.

Schwenke, W. 1978. Die Forstschädlinge Europas. Band 3, Schmetterlinge, Verlag Paul Parey, Hamburg und Berlin.

Schwenke, W. 1982. Die Forstschädlinge Europas. Band 4, Hautflügler und Zweiflügler, Verlag Paul Parey, Hamburg und Berlin.

Schwerdtfeger, F. 1981. Die Waldkrankheiten. 4. edition, Parey, Hamburg und Berlin.

Seemann, D., Bouffier, V., Kehr, R., Wulf, A. and Schröder, T. 2001. Die Esskastanie (*Castanea sativa* Mill.) in Deutschland und ihre Gefährdung durch den Kastanienrindenkrebs (*Cryphonectria parasitica* [Murr.] Barr). Nachrichtenblatt des Deutschen Pflanzenschutzdienstes 53: 49–60.

Seemann, D. and Unger, J. 1993. Rindenkrebs der Eßkastanie in der Bundesrepublik Deutschland. Nachrichtenblatt des Deutschen Pflanzenschutzdienstes 45: 120–122.

Seemann, D. and Zajonc, J. 1994. Rindenkrebs der Eßkastanie (*Cryphonectria parasitica*) in Südwestdeutschland. European journal of forest pathology 24: 241–244.

Shaw, C.G. and Kile, G.A. 1991. Armillaria root disease. Forest Service Agriculture Handbook No. 691, Washington, D.C.

Sinclair, W.A., Lyon, H.H. and Johnson, W.T. 1993. Disease of trees and shrubs. 2. edition, Cornell University Press, Ithaca and London.

Stipes, R.J. and Campana, R.J. 1981. Compendium of Elm Diseases. American Phytopathological Society Press, St. Paul., Minn.Werres, S. 1996. *Phytophthora* spp. an Erle Eiche Kastanie und anderen Bäumen – aktueller Stand in der Bundesrepublik Deutschland. Nachrichtenblatt des Deutschen Pflanzenschutzdienstes 48: 160.

Strouts, R.G. and Winter, T.G. 2000. Diagnosis of ill-health in trees. 2. ed., Forestry Commission.

Tainter, F.H. and Baker, F.A. 1996. Principles of Forest Pathology. John Wiley & Sons, Inc., New York.

Werres, S. 2000. Neuartiges Erlensterben durch *Phytophthora*-Arten. Jahrbuch der Baumpflege, Thalacker-Verlag, Braunschweig. Pp. 212–216.

Werres, S. 2001. Absterbeerscheinungen an Rhododendron und Viburnum – neue Phytophthora-Art entdeckt. Deutsche Baumschule 4: 40–41.

Werres, S. and Hahn, R. 2000. *Phytophthora* spp. on trees – present situation in Germany. International Symposium on plant health in urban horticulture. Mitteilungen aus der Biologischen Bundesanstalt für Land- und Forstwirtschaft 370: 78–80.

Werres, S. and Marwitz, R. 1997. Triebsterben an Rhododendron – Unbekannte *Phytophthora*, Deutscher Gartenbau 21/97: 1166–1168.

Werres, S., Marwitz, R., Man In'T Veld, W.A., De Cock, A.W.A.M., Bonants, P.J.M., De Weerdt, Themann, K., Ilieva, E. and Baayen, R.B. 2001. *Phytophthora ramorum* sp. nov., a new pathogen on Rhododendron and Viburnum. Mycological Research 105: 1166–1175.

Wulf, A. 1988. *Pleuroceras pseudoplatani* (v. Tubeuf) Monod, Erreger einer Blattbräune an Bergahorn (*Acer pseudoplatani* L.). Nachrichtenblatt des Deutschen Pflanzenschutzdienstes 40(5): 65–70.

Wulf, A. 1990. Über die Bedeutung von *Diplodina acerina* (Pass.) Sutton und anderen Blattpilzen als Antagonisten der Fenstergallmücke Dasineura vitrina Kffr. An Bergahorn (*Acer pseudoplatanus* L.). Nachrichtblatt des Deutschen Pflanzenschutzdienstes 42(7): 97–102.

Wulf, A. 1994. Pilzbedingte Blattkrankheiten an Ahorn. Schriften aus der Forstlichen Fakultät der Universität Göttingen und der Niedersächsischen Forstlichen Versuchsanstalt, Band 116, Sauerländer,s Verlag Frankfurt am Main.

Wulf, A. 1999. Zur Verschleppung des Asiatischen Laubholz-Bockkäfers Anoplophora glabripennis nach Nordamerika und über sein Gefährdungspotential für die europäischen Wald-, Park- und Straßenbäume. Nachrichtenblatt des Deutschen Pflanzenschutzdienstes 51(3): 53–57.

Wulf, A. and Pehl, L. 2002. Asiatischer Laubholzbockkäfer, information leaflet, BBA

Wulf and Kehr: Protection of Valuable Broadleaved Stands: Diseases, pests and natural disturbances. Plate I

Maple (Acer pseudoplatanus): Fig. 1: Powdery mildew caused by *Sawadea tulasnei.* Fig. 2: Tar spot disease caused by *Rhytisma acerinum.* Fig. 3: Coral spot caused by *Nectria cinnabarina,* sporodochia of the *Tubercularia anamorph* shown here. Fig. 4: Discolorations in sapwood caused by *Verticillium* wilt. Fig. 5: Holes in leaves caused by window gall *Dasineura vitrina.* Fig. 6: Leaf feeding by caterpilar of Sycamore Moth, *Acronicta aceris.*

Wulf and Kehr: Protection of Valuable Broadleaved Stands: Diseases, pests and natural disturbances. Plate II

Ash (Fraxinus excelsior): Fig. 1: Damage by rabbits. Fig. 2: Bark stripping by Red Deer. Fig. 3: Bark damage by Hornets (Vespa crabo). Fig. 4: Bacterial canker caused by *Pseudomonas syringae* subsp. *savastanoi* pv. *fraxini*. Fig. 5: Fungal target canker caused by *Nectria galligena*. Fig. 6: Feeding damage by Lilac leaf miner (*Caloptilia syringella*)

Wulf and Kehr: Protection of Valuable Broadleaved Stands: Diseases, pests and natural disturbances. Plate III

Diseases and disorders of various broadleaved trees: Fig. 1: Dutch Elm disease, healthy (right) and dying (left) elm and characteristic stains in sapwood (Inset). Fig. 2: Chestnut (*Castanea sativa*) canker caused by Cryphonectria parasitica. Note the dark advancing canker margin. Fig. 3: Slime flux following death of cambium in Alder (*Alnus fruticosa*) caused by Phytophthora sp. Fig. 4: Gummosis following Monilia infection in Sweet Cherry (*Prunus avium*)

4. Management of Valuable Broadleaves

4.1 Crown Architecture of Valuable Broadleaved Species

Daniel Barthélémy[1], Yves Caraglio[2] and Sylvie Sabatier[2]
[1]INRA, UMR AMAP, Montpellier, France
[2]CIRAD, UMR AMAP, Montpellier, France

Abstract

The architecture and growth dynamics of *Prunus avium* L., *Fraxinus excelsior* L. and *Acer pseudoplatanus* L. were studied in several regions of France. Trees growing in open field and in the understory of forest stands were observed. Numerous trees were studied at several stages of development in order to identify the morphological and architectural features of these species. The establishment of the architectural unit and the strategy of crown development were described for each species. The architectural bases of interspecific and intraspecific variations are discussed and their consequences on understanding of forest structure and plantation management are suggested.

Keywords: crown architecture, crown development, valuable broadleaves, plantation management

4.1.1 Introduction

Initiated in tropical areas, the architectural approach (Barthélémy et al. 1991, 1989 Hallé et al. 1978, Hallé and Oldeman 1970) has provided a powerful tool for studying plant form. Investigations in this field have quickly spread to the study of temperate tree species among which high quality hardwood broadleaved species such as ash (Barthélémy and Caraglio 2007), beech (Nicolini 1997), wild cherry (Caraglio 1996), chestnut (Caraglio 1992), plane (Caraglio and Edelin 1990), sycamore (Gardère 1995) and walnut (Barthélémy et al. 1997b, 1995) may be cited. Architectural studies of ash, wild cherry and sycamore species started some years ago in the context of an agreement between I.D.F. (Institut pour le Développement Forestier) and C.I.R.A.D. (Centre de coopération Internationale en Recherche Agronomique pour le Développement). These studies of temperate species led to a general interpretation of their architectural sequence of development and were then extended to the analysis of inter- and intraspecific architectural variations. In this paper the architectural sequences of the development of *Prunus avium*, *Fraxinus excelsior* and *Acer pseudoplatanus* are described. Intra- and interspecific variations are analysed for these species and some additional remarks are given for chestnut and walnut trees. Finally, the influence of genetic and environmental factors on the observed architectural variations are discussed.

4.1.2 Material and methods

Material

Trees analysed for the identification of the architectural sequence and for the study of intra- and interspecific variations, which were growing in plantation and/or in sub-spontaneous regeneration, in open field or in natural forest environment, were observed in several regions of France.

Architectural analysis

The architecture of a plant depends on the nature and relative arrangement of each of its parts and is, at any given time, the expression of an equilibrium between endogenous growth processes and exogenous constraints exerted by the environment. The aim of the architectural analysis is to reveal and identify these endogenous processes and to understand the influence of exogenous factors by means of observation. Considering the plant as a whole from its germination to its death, architectural analysis is essentially a dynamic and holistic approach to understanding plant development. For each species and each stage of development observations are made on varying numbers of trees depending on the complexity of the architecture. When the plant is small there is no particular observational problem and it can therefore be observed directly. For tall adult trees, however, observations are carried out on standing trees from ground level with the use of binoculars. This technique is efficient when the juvenile architecture is already known and allows a skilled observer to recognise, even in a high crown, very precise structural details that would otherwise pass unnoticed. This technique must be confirmed with checks on dead fallen branches and by direct observation of fallen or harvested trees. These observations allow for the characterization of the growth strategy, branching pattern and sequence of development of any species. The results are summarised in a set of diagrams and tables that allow the architecture of any given individual of the species to be better understood.

Growth pattern

In species originating from temperate areas annual stem growth takes place in one or more successive extension phases. The portion of leafy axis established during a continuous extension phase corresponds to a growth unit (Hallé et al. 1978). The portion of a leafy axis built up over a single growth year corresponds to an annual shoot and may consist of one or more growth units (Barthélémy and Caraglio 2007). Morphological markers can be used to reconstitute tree growth *a posteriori*. On young axes inter-annual or winter halts in growth are clearly shown by the presence of several cataphylls (*e.g.* scale leaves) associated with very short internodes. On the oldest parts of the axes the annual shoot limits are marked off by sets of scars left by the cataphylls that have fallen down. Leaf scars usually remain visible on the bark for many years thus allowing for the identification of the limits between shoots and the retrospective study of tree growth.

4.1.3 Architectural sequence of development

The aim of the developmental sequence study is to reveal the specific growth and branching strategies in order to obtain an endogenous developmental reference for the further analyse of architectural intra- and interspecific variations.

Wild cherry: *Prunus avium* L. *Rosaceae*
In natural conditions the germination of wild cherry is rare because the kernel is sensitive to attack by fungi and weevils. In nursery conditions seedlings present an unbranched stem bearing, above pulpy cotyledons, spiral alternate entire leaves with two stipules at the base of the petiole. Each year the apical meristem produces an axis portion which, from the base to the top, bears in succession a series of cataphylls, abortive leaves, green leaves with axillary buds and finally some small leaves without visible axillary buds. In wild cherry cataphyll scars marking the limit between two successive annual shoots (Figure 4.1a) may remain visible on the bark of axes for as long as 10 to 20 years. On a young tree annual shoots normally appear in one single flush of growth, although in good growth conditions two, or more rarely three, successive flushes of growth (respectively named “spring shoot”, “shoot of Saint-John” and “August shoot” according to their time of occurrence) may be observed.

Auxiliary buds give rise to long shoots in their upper part and to short shoots in their lower part (Figure 4.1b) the year following the development of an annual shoot. As for most forest species native from temperate areas, branching is thus delayed by one year (i.e. proleptic). As the tree grows the annual shoots of the main stem bear bigger and more ramified branches in a distal position whereas short lateral axes develop in a proximal position. In open field first flowering occurs on three to four-year-old trees. Inflorescences are borne on short shoots and the most proximal part of one-year-old long shoots (Figure 4.1c). Each leaf node of a short shoot may bear a lateral inflorescence while only the first leaves of long shoots do so.

In all mature wild cherry trees four different categories of axes may be recognised. The main stem is an orthotropic monopodial axis with indefinite and rhythmic growth; its branching pattern is rhythmic. Branches are horizontal or slanted with long-term definite and rhythmic growth and monopodial and rhythmic branching. Branchlets present a horizontal growth direction, have a medium-term definite and rhythmic growth and bear only brachyblasts. Brachyblasts are very short axes producing a single cluster of leaves each year; they do not have a precise growth direction; they experience short-term determinate growth (5 to 7 years) and remain unbranched. This precise organisation concerning the kind and relative position of morphologically discernible categories of axes corresponds to the expression of the specific functional elementary architecture, also called “architectural unit” (Figure 4.2a).

The tree continues its development according to this pattern during numerous years and may reach a height of 20 to 25 meters in forest conditions, with a small

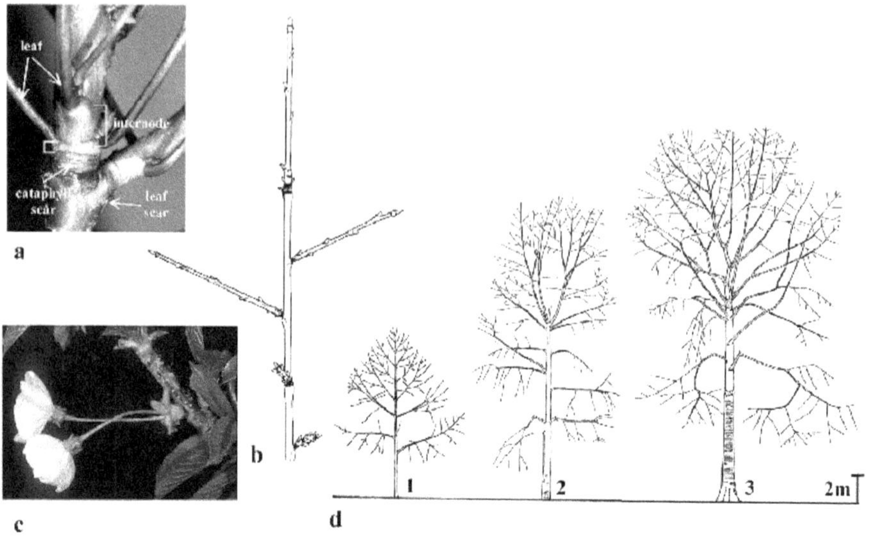

Figure 4.1. Architecture of *Prunus avium* (from Caraglio 1996). **a**. Morphological markers resulting from tree growth. Scars of cataphylls associated with short internodes indicate the winter halt in growth. **b**. Structure of a two-year-old stem portion. **c**. Lateral inflorescence. **d**. Young (**1**), adult (**2**) and old trees (**3**) in forest environment.

crown similar to that of a young tree, and still conform to its architectural unit (Figure 4.2b2). As the tree becomes older some branches located in the upper part of the tree become larger and similar in structure to the main stem (Figures 4.1d2 and 4.2b3). This morphological transformation results from a duplication of the initial architectural unit and corresponds to following a "reiteration" (Hallé et al. 1978) process that is involved in the mature tree crown construction according to an "architectural metamorphosis" (Edelin 1984, Hallé and Ng 1981). At this stage the tree reaches its maximal height (Figure 4.2b4). The process of duplication continues and the crown of an old tree is built up by numerous forks consisting of ultimate axes made of short shoots (Figure 4.1d3). At this time the sagging of limbs of the crown results in the formation of the tabular-shape crown. After 18 to 20 years these axes die. New structures then take place from latent buds located on the well-developed branches and appear progressively closer to the point of branch insertion as the tree ages. Floriferous structures may derive from buds associated to those scars which are limiting the shoots. This ultimate process determinates the senescent phase of the tree.

Common ash: *Fraxinus excelsior* L. *Oleaceae*

In forest conditions ash germinates in the understory. One-year-old seedlings develop an erect and unbranched stem which produces, above the cotyledons, two opposite entire and denticulate leaves (Figure 4.3a). During the first years of growth the apical meristem of the main stem forms annual shoots which increase yearly.

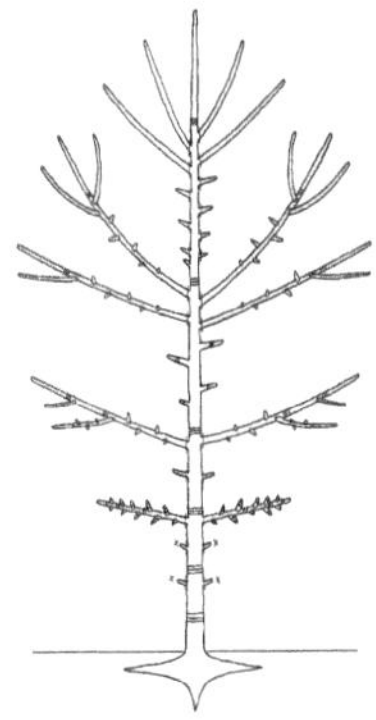

a

Main stem	Branches	Branchlets	Brachyblasts
Indeterminate growth	Long term determinate growth	Medium term determinate growth	Short term determinate growth
Vertical growth direction	Horizontal to slanted growth direction	Horizontal growth direction	No precise growth direction
First leaves of GU bear inflorescence	First leaves of GU bear inflorescence	First leaves of GU bear inflorescence	All leaves of GU bear inflorescence
8 to 50 leaves	8 to 30 leaves	10 leaves	3 to 7 leaves
Bears 4 to 6 long shoots	Bear 2 to 4 long shoots	no	no

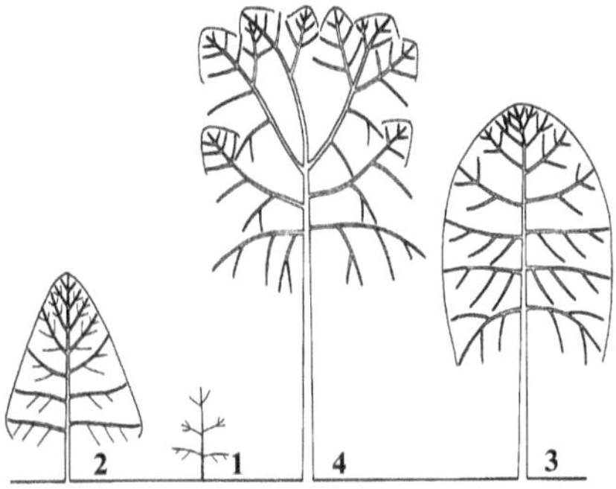

b

Figure 4.2. Architecture of *Prunus avium* (from Caraglio 1996). **a**. Architectural unit. The table summaries the morphological traits of all axis categories. The diagrammatic representation of the tree represents the relative position of the main axis categories. GU: growth unit. **b**. Developmental sequence in forest environment. **1**: young vegetative tree; **2**: young tree conforms to its elementary architecture; **3**: construction of the mature crown which becomes rounder; **4**: adult tree exhibiting a tabular-shape crown. The ultimate axes are not noted.

These shoots consist of one or two pairs of cataphylls followed by two to three pairs of compound green leaves composed of three, five, and seven leaflets in subsequent years. Branching occurs when the leaves of the main stem possess nine to fifteen leaflets (Figures 4.3b–d). The first branches consist of a succession of unbranched short shoots which bear one to two proximal cataphyll pairs and one to two compound leaves of five to seven leaflets (Figure 4.3e).

In the lower part of the older tree the branches consist of a succession of short annual shoots composed of one or two pairs of green leaves. In the upper part of the tree the annual shoots of the main stem are bigger and possess six to nine pairs of leaves. They give rise to branches whose sizes decrease from the extremity to the base of each parental bearing shoot. Branch shoots are long and consist of five to seven pairs of leaves. They bear branchlets which possess a structure similar to the first lateral axes formed in the young tree (Figure 4.3f).

As the tree continues its development the annual shoots of the main stem become regularly branched and bear bigger and more ramified branches in distal position

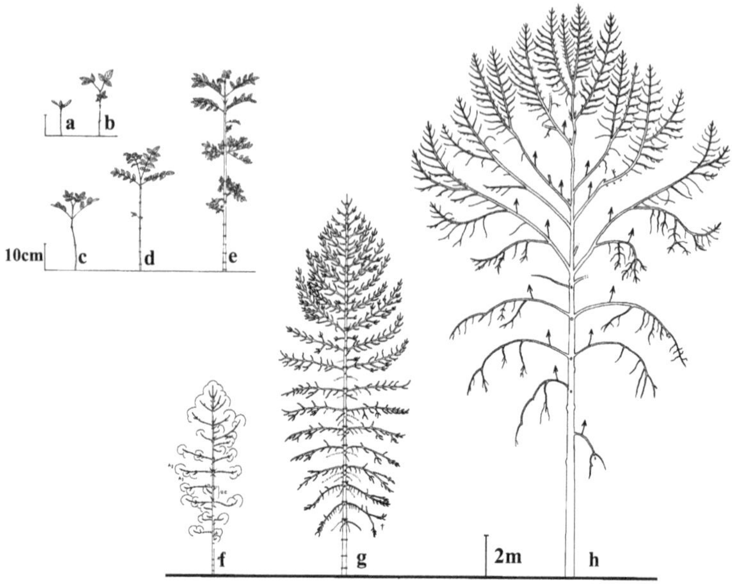

Figure 4.3. Architectural development of *Fraxinus excelsior*. **a** to **e**: young trees in forest understory; **f**: establishment of the first branches; **g**: death and self-pruning of lower branches in an young ash with a structure conforming to the architectural unit; **h**: construction of the crown on the adult tree (from Barthélémy and Caraglio 2007). (#) winter halt of growth.

whereas short lateral axes similar to branchlets develop in proximal position. At this step the lower branches progressively die and self-prune whereas the upper branches are abundantly branched. The adult tree has a cone-shaped crown and a simple architecture organised around the main stem (Figure 4.3g). It expresses an architectural unit made of three axis categories: main stem, branches and branchlets (Figure 4.4a). First flowering generally occurs at this developmental stage and inflorescences may develop laterally on all axis categories.

As the tree grows branching is more abundant and branches become bigger. The upper branches turn up and their structure progressively reproduces the architecture of the young tree conforming to its architectural unit, thus forming reiterated complexes which become the limbs of the older tree (Figure 4.3h). During crown construction this phenomenon of fork formation reproduces itself many times. The size of structures between two successive forks progressively decreases. On an old tree that is reaching its maximal size and development, the upper part of the crown consists of small and homogeneous structures (Figure 4.4b4) whereas in the lower part of the tree the oldest branches die and are finally shed. In the following stages reiterated complexes are initiated from axillary latent buds located near the damaged part and on the upper side of initial branches.

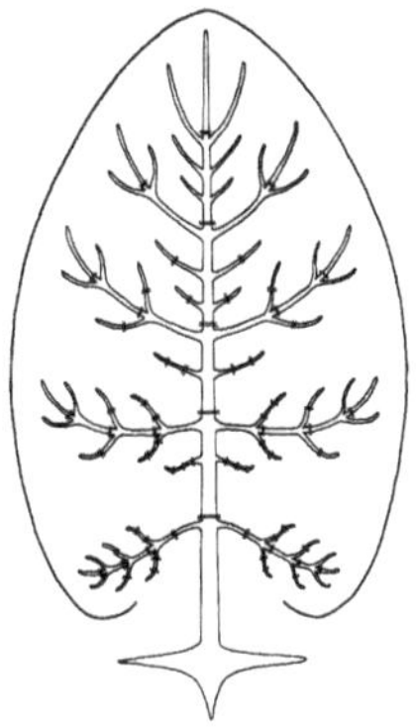

a

Main stem	Branches	Branchlets
Indeterminate growth	Long term determinate growth	Medium term determinate growth
Vertical growth direction	Horizontal to slanted growth direction	Horizontal growth direction
Bears inflorescence	Bear inflorescence	Bear inflorescence
Long GU	GU with mean size	Short GU
Very branched	Few branched	Unbranched

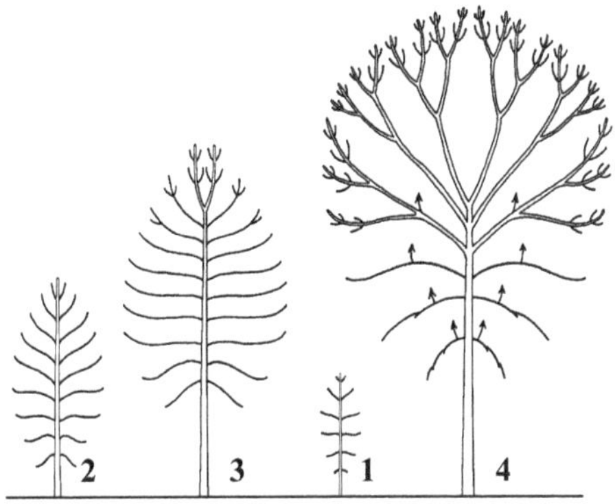

b

Figure 4.4. Architecture of *Fraxinus excelsior*. **a**. Architectural unit. The table summaries the morphological traits of all axis categories. The diagrammatic representation of the tree represents the relative position of the main axis categories (from Barthélémy and Caraglio 2007). GU: growth unit. **b**. Developmental sequence in forest environment. **1**: young vegetative tree; **2**: young tree conforming to its elementary architecture; **3**: adult tree and construction of the crown which becomes rounder; **4**: adult tree and decreasing size of successive duplicated structures in the crown. Ultimate axes are not represented (from Barthélémy et al. 1995).

Sycamore: *Acer pseudoplatanus* Aceraceae

In forest conditions one-year-old sycamore seedlings present an unbranched axis consisting, above the cotyledons, of two opposite entire leaves (Figure 4.5a). During the first years of growth annual shoots with increasing size are successively produced. These shoots consist of five pairs of cataphylls followed by one to three pairs of green lobate leaves. First branches consist of a succession of unbranched short shoots (Figure 4.6a). As the tree grows new annual shoots of the main stem progressively possess four to nine pairs of leaves and give rise to branches which display decreasing sizes from the extremity to the basis of the parental shoots. The bigger branches consist of long shoots with three to four pairs of leaves and bear branchlets which are constituted of short annual shoots bearing one or two pairs of leaves. Their structure is similar to the first lateral axes formed in the youngest tree (Figure 4.6b).

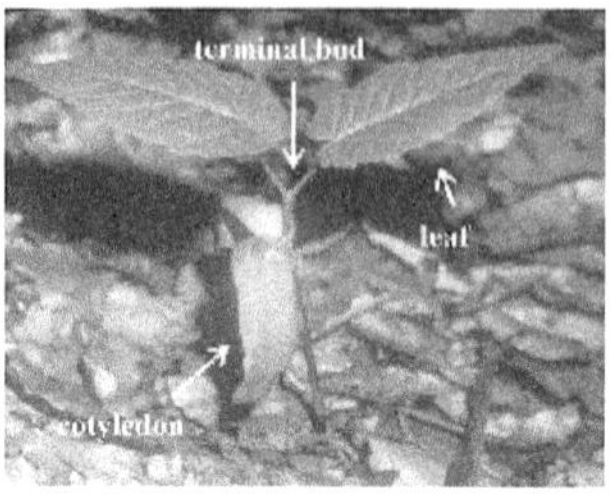

a

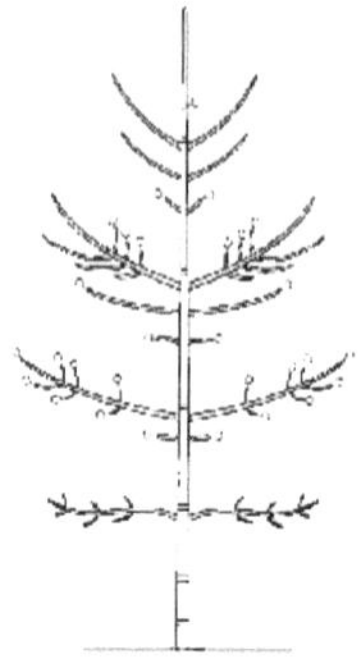

c

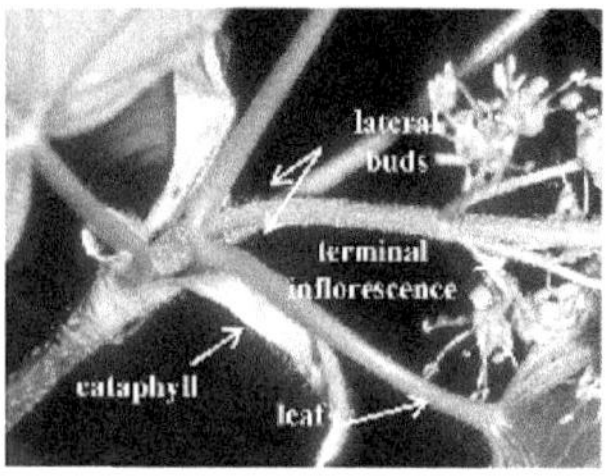

b

Main stem	Branches	Branchlets
Long term determinate growth	Medium term determinate growth	Short term determinate growth
Vertical growth direction	Horizontal to slanted growth direction	Horizontal growth direction
Bears inflorescence	Bear inflorescence	Bear inflorescence
Long GU	GU with medium size	Short GU
Very branched	Few branched	Unbranched

Figure 4.5. Architecture of *Acer pseudoplatanus*. **a**. One-year-old seedling. **b**. Terminal flowering on a spring shoot. **c**. Architectural unit. The table summarizes the morphological traits of all axis categories. The diagrammatic representation of the tree represents the relative position of the main axis categories. GU: growth unit.

Between the ages of 15–20 years trees become sexually mature and flowering is terminal on spring shoots (Figure 4.5b). After the development of a terminal inflorescence the flowering axis continues its growth by the development of one or two relay axes which derive from lateral buds localised just under the inflorescence. At this time the initially monopodial developmental pattern becomes sympodial. First flowering usually only affects the ultimate axes located at crown periphery (Figure 4.6c). During the following years flowering progressively invades the whole crown of the trees and the terminal flowering on the main stem involves the development of relay axes at its extremity. Upon first flowering on the main stem one dominant relay axis normally develops in its prolongation thereby allowing it to maintain a rectilinear aspect. This relay axis does not produce inflorescence during the two or three years before its next flowering.

At this stage of development the tree has a vertical sympodial main stem bearing tiers of horizontal ramified branches. The tree expresses its architectural unit (Figure 4.5c) and has a cone-shape crown (Figure 4.6c).

As the tree grows the upper branches of the crown turn up and their morphological traits are similar to those of the main stem. This process of crown construction is

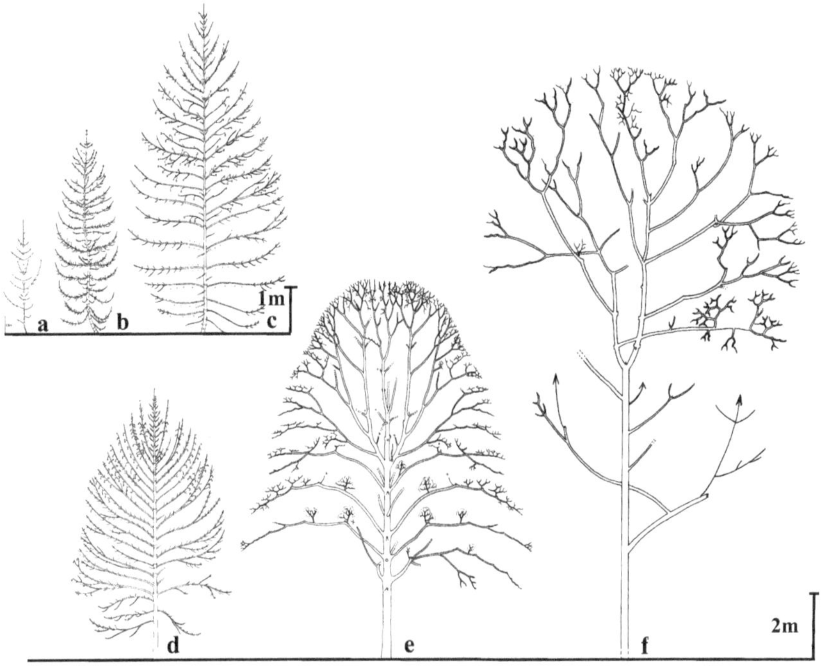

Figure 4.6. Architectural development of *Acer pseudoplatanus* in mountain conditions. **a** and **b**: vegetative young tree and establishment of the first branches; **c**: first flowering; **d**: adult tree and crown construction by metamorphosis of the upper branches; **e**: adult tree and decreasing of the size of structures between two successive forks; **f**: older tree and death of lowest main branches.

associated with a modification of the crown shape which becomes more spherical (Figure 4.6d). At a given stage of development of the tree the dominance of one relay axis formed after terminal flowering naturally disappears and a fork is produced at the top of the main stem. This phenomenon of fork formation reproduces itself during the successive flowering phases and the size of the structures produced between two forks decreases progressively during tree ageing (Figure 4.6e). The crown of the old tree is thus entirely built up by numerous small forks whereas, in its lower part, the oldest branches die and are shed (Figure 4.6f). At this time reiterated complexes, exhibiting a structure similar to that of young tree, may arise from the development of latent buds on the intact part of branches.

4.1.4 Architectural variations

Interspecific variations in the number, nature and relative position of the categories of axes of the architectural unit were found between wild cherry, ash and sycamore.

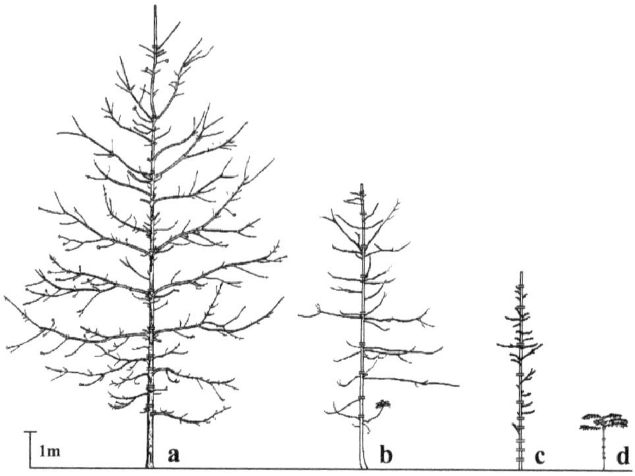

Figure 4.7. Architectural variations according to light availability. Variations in the architecture of young trees of *Juglans nigra* according to increasing (left to right) canopy density. **a**: flowering 15-year-old trees growing in open field; **b**: vegetative 13-year-old tree growing below a clear canopy; **c**: vegetative 15-years-old tree in the understory of a dense canopy; **d**: vegetative unbranched 6-year-old tree growing in a dark understory (from Barthélémy et al. 1997b).

Only wild cherry trees possess brachyblasts which are a very specialised category of axes. The architectural unit is characterised by a more hierarchical architecture and a stronger differentiation of the categories of axes for wild cherry than for both other species. Branching is monopodial and flowering is lateral in wild cherry and ash, whereas the initial monopodial pattern turns to sympodial on mature terminal flowering sycamores. In spite of these differences, during the first years of flowering the main stem of sycamore is made up of a rectilinear sympodial succession of dominant relay axes. On old trees interspecific differences in crown physiognomy are observed: wild cherry exhibits a flat crown whereas the crown is rounded in ash and sycamore.

The duration of the establishment sequence of a particular stage of development and the exact structure presented by a tree at this particular stage may be subject to great quantitative variations and are modulated by environmental factors such as light availability and population density. When trees, of all species, are growing in extreme conditions of light deficiency, as in the understory of a closed canopy, they remain very small, vegetative and unbranched and develop only short annual shoots during many years. In the case of sycamore each shoot consists of two pairs of green leaves; this phase of slow growth may last up to 20 years (Gardère 1995). Similar processes are observed on walnut species (Barthélémy et al. 1997b, 1995): (*i*) under conditions of light deficiency *Juglans nigra* exhibits an unbranched main stem composed of successive short annual shoots (Figure 4.7d). For individuals of this species more or less the same age, the size of both the main stem and the branches increases with an increase in light availability (Figures 4.7 b–c). First

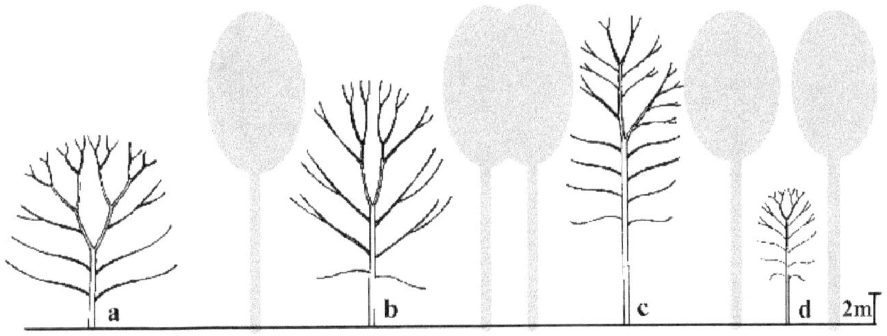

Figure 4.8. Architectural variations according to light availability. Variations in crown physiognomy with population density in adult *Juglans regia*. The height of the first fork, which derives from terminal flowering of the main stem and determines the size of the butt log, increases from full sun (**a**) to forest stand conditions (**b** and **c**). When the tree remains below a canopy, its architectural development is reduced (**d**).

flowering also appears later on trees growing in the understory than on those growing in full light (Figure 4.7a); (*ii*) for *Juglans regia* of an older age, trees which grow in a forest gap or with a lateral shelter exhibit a taller but thinner stem as well as shorter and thinner branches than open-grow trees. Self-pruning is also more intensive and the trees fork later and higher. Thus the size of their butt log is greater than that of open-grow trees (Figures 4.8a–c). In the understory of a very dense canopy growth becomes very slow and trees exhibit a reduced structure (Figure 4.8d).

As in the case of walnut, environmental conditions mostly affect the branch to stem length ratio of wild cherry as well as the time and height at which the main crown limbs appear (Caraglio 1996; Figure 4.9). When light availability is reduced the growth rate of all axes is low and wild cherry exhibits a tabular-shape crown much earlier. In the same environmental conditions a young *Castanea sativa* also exhibits a tabular-shape structure due to the absence of dominance of the main axis. As the tree grows the buds located at the tree base give rise to new axes which have a structure similar to that of the parent axis (Caraglio 1992; Figure 4.10a). Small trees in the understory may survive for several years with a reduced structure which may be referred to as "waiting form" or "survival form"(Grosfeld et al. 1999, Nicolini 1997, Caraglio 1992). These trees can further develop only if there is an opening in the above canopy; in such a case a vertical relay axis appears which is composed of long growth units and gives rise to the main stem (Figures 4.10b–d). On older trees an abrupt change in tree environment may lead to the appearance of epicormic branches on the stem (Nicolini et al. 2001, Nicolini and Caraglio 1994).

In the latter stages, trees growing in forest conditions finally exhibit a crown form which resembles that of open-grown trees. However, they are taller, their butt log is longer and they reach this stage of development at an older age (see Figures 4.8 and 4.9). In contrast, as in the case of open-grown trees, trees growing on poor soils are smaller and exhibit a spherical-shape crown at a younger age (Barthélémy et al. 1997b).

Figure 4.9. Architectural variations according to light availability. Variations in wild cherry crown construction between a young individual growing in open field (**a**), an individual growing in open forest conditions (**b**) and individual growing in dense forest conditions (**c**; from Caraglio 1996).

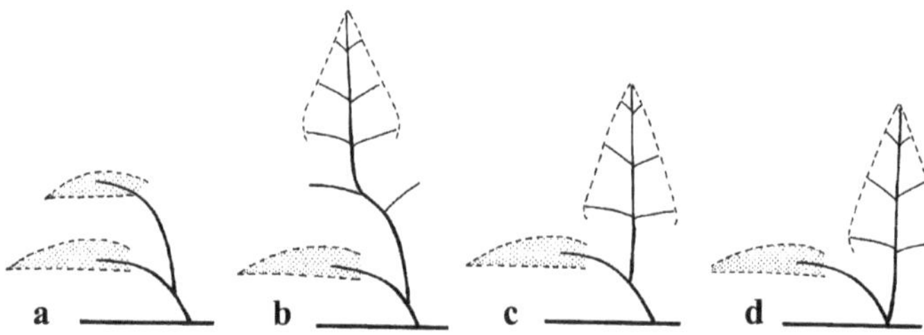

Figure 4.10. Architectural variations according to light availability. "Waiting forms" of *Castanea sativa* growing in the shaded understory of a forest (from Caraglio 1992).

All these results comply with other studies of the plasticity of plant architecture which together show that the sequence of development of a species is the expression of an endogenous process which is quantitatively modulated by environmental factors (Nicolini et al. 2001, Grosfeld et al. 1999, Barthélémy and Caraglio 2007, Nicolini and Caraglio, 1994, Hallé et al. 1978).

The size and the insertion angle of branches, which are directly affected by environmental factors, may also vary among progenies and clones of different populations of wild cherry (Lemoine et al. 1992).

Figure 4.11. In *Prunus avium*, pattern of main stem construction after death of its shoot apex (from Caraglio 1996).

In all these species after an apex death the growth of the main stem continues through the development of a relay shoot from one of the most distal axillary latent buds. The main stem is then constituted of a succession of relay axes which very often leads to more or less marked form defects (Figure 4.11).

These defects usually persist for a longer period for wild cherry and ash than for sycamore and walnut. Thus the early pruning during the growing season may improve the rectilinear aspect of the main stem. For walnut, formative pruning (*i.e.* fork removal by cutting the co-dominant axis opposite to the selected relay axis) was performed on young trees at different periods during the growing season: March, May, June and July (Sabatier et al. 2000). In terms of establishment and straightening of the unique relay axis, the best results are obtained when pruning is performed in June or July.

With the aim of analysing the influence of artificial pruning, studies of the architectural response of young planted wild cherry trees were carried out (Caraglio et al. 2000): unpruned control trees were compared to systematically pruned trees (i.e. removal of all branches except the last pseudo-whorl of branches) and selectively pruned trees (i.e. removal of the biggest branches of the pseudo-whorl). Pruning was applied in March during three successive years. Systematic pruning has both immediate and delayed effects: in the year of pruning the remaining branches are thicker and leaves are larger; during the first years after pruning stem shoots are longer and the stem has a more rectilinear aspect; the frequency of epicormic branches gradually increases during the successive years and affects the butt log quality. Similar morphological responses have been observed in young walnut trees after a systematic pruning applied in July (Sabatier et al. 2000). A practical consequence of these results is the necessity to define the optimal pruning height. In wild cherry selective pruning leads to an increase of the size of the remaining branches (Caraglio et al. 2000). Since dominantly forest grown wild cherry and walnut trees have more horizontal and thinner branches, pruning is easier and more efficient for these trees than for open-grown trees.

4.1.5 Conclusions

The main architectural differences among the studied species concern the degree of differentiation between their categories of axes, the pattern of development of the main stem, the pattern of crown construction and the resulting crown geometry. Even if they are not modified in their qualitative aspects, these features may be strongly affected by environmental factors. For example, the first main fork appears earlier and at a lower height when trees grow in full light or on poor sites (i.e. sites with a low fertility). The height of the butt log thus depends very much on environmental conditions. These phenomena have consequences on timber quality; for example, the trees which grow in a forest gap or with a lateral shelter have a longer butt log, thinner branches and less stem defects.

These studies have allowed us to define the main qualitative architectural features of some valuable broadleaved species as well as their variations across environmental conditions. Knowledge of architectural diversity and plasticity provides valuable information which can be used to identify the physiological status of trees, to improve forest and plantation management (for example, to obtain a good timber quality), or in breeding and genetic improvement programs.

References

Barthélémy D. and Caraglio, Y. 2007 Plant architecture: a dynamic, multilevel and comprehensive approach to plant form, structure and ontogeny. Annals of Botany 99:375–407.

Barthélémy, D., Edelin, C. and Hallé, F. 1989 Architectural concepts for tropical trees. In: Tropical Forests. Botanical dynamics, speciation and diversity. Holm-Nieslen L.B. and Balsler H. (eds.) Academic Press, London. Pp 89-100.

Barthélémy, D., Edelin, C. and Hallé, F. 1991 Canopy architecture. In: Physiology of trees. Raghavendra, A.S. (ed.) John Wiley and Sons Inc. Pp 1–20.

Barthélémy, D., Sabatier, S. and Pascal, O. 1995. Le développement architectural du noyer commun *Juglans regia* L. (Juglandaceae). Forêt-Entreprise 103: 61–68.

Barthélémy, D., Sabatier, S. and Pascal, O. 1997b. Le développement architectural du noyer noir *Juglans nigra* L. (Juglandaceae). Forêt-Entreprise 115: 40–47.

Caraglio, Y. 1992 Naissance d'un arbre. In: Le chataignier, un arbre, un bois. Bourgeois C., Editions IDF, Paris.

Caraglio, Y. 1996. Le développement architectural du merisier. Forêt Entreprise 107: 72–80.

Caraglio, Y., Becquey, J., Gallois, F. and Vidal Ch. 2000. Réaction de jeunes merisiers à la taille en boisement de terre agricole. Forêt-Entreprise 13: 25–29.

Caraglio, Y. and Edelin, C. 1990. Architecture et dynamique de la croissance du platane, *Platanus hybrida* Brot. (Platanaceae). Bulletin de la Société Botanique de France, Lettres Botaniques 137 (4–5): 279–291.

Edelin, C. 1984 L'architecture monopodiale: l'exemple de quelques arbres d'Asie tropicale. Thèse, Université Montpellier II. 258 p.

Gardère, I. 1995 Influence de l'intensité du couvert sur le développement architectural de jeunes Erables sycomores *Acer pseudoplatanus* (Aceraceae). Mémoire de DEA de Biologie Forestière, Université Nancy I. 19 p.

Grosfeld, J., Barthélémy, D. and Brion, C. 1999 Architectural variations of *Araucaria araucana* (Molina) K. Koch (Araucaraceae) in its natural habitat. In: The evolution of Plant Architecture. Kurmann, M.H. and Hemsley, A.R. (eds.) Royal Botanic Gardens, Kew, London.

Hallé, F. and Ng, F.S.P. 1981. Crown construction in mature Dipterocarp trees. Malaysian Forestry 44 (2–3): 222–233.

Hallé, F., Oldeman, R.A.A. 1970 Essai sur l'architecture et la dynamique de croissance des arbres tropicaux. Masson et cie, Paris. 178 p.

Hallé, F., Oldeman, R.A.A. and Tomlinson, P.B. 1978 Tropical trees and forests. An architectural analysis. Springer Verlag, Berlin. 441 p.

Lemoine, M., Dufour, J. and Santi, F. 1992 Le merisier. In Amélioration, des espèces végétales cultivées, objectifs et critères de sélection. Gallais, A. and Bannerot, H. (eds.) INRA Editions, Paris.

Nicolini, E. 1997 Approche morphologique du développement du hêtre (*Fagus sylvatica* L.). PhD thesis, University of Montpellier 2, France.

Nicolini, E. and Caraglio Y. 1994. L'influence de divers caractères architecturaux sur l'apparition de la fourche chez *Fagus sylvatica* L., en fonction de l'absence ou la présence d'un couvert. Canadian Journal of Botany 75: 1723–1734.

Nicolini, E., Chanson, B. and Bonne, F. 2001. Stem and epicormic branch formation in understorey beech trees (*Fagus sylvatica* L.). Annals of Botany 87: 737–750.

Sabatier, S., Barthélémy, D., Becquey, J. and Perrier, S. 2000. Taille et architecture chez de jeunes noyers hybrides. Forêt-Entreprise 132: 54–58.

4.2 Modeling Natural Pruning of Common Ash, Sycamore and Wild Cherry

Sebastian Hein
Institute for Forest Growth, University of Freiburg, Germany

Abstract

Branches and knots can greatly affect wood quality. As no models for controlling wood quality of common ash (*Fraxinus excelsior* L.), sycamore (*Acer pseudoplatanus* L.,) and wild cherry (*Prunus avium* L.) were readily available in the literature, a model was established to predict the length of the branch free bole. The model included the total height of the tree, the diameter at breast height, age of the tree and a parameter for forked and non-forked trees. With the aid of stem form functions, the diameter of the knotty core was predicted depending on site quality and competition. Decision rules were established to control the length of the branch free bole and the width of the knotty core. Also presented in this chapter is a two phase system for controlling natural pruning dynamics.

Keywords: crop trees, natural pruning, artificial pruning, Fraxinus excelsior L., Acer pseudoplatanus L., Prunus avium L.

4.2.1 Producing valuable timber

Branchiness

For the production of valuable timber the quality of the log is of great importance. Timber quality includes many characteristics, but since branches and knots affect the mechanical, chemical and especially the aesthetic properties of round wood and sawn timber, they can be regarded as main characteristics to controlling quality. Whereas there are some investigations on the quality of common ash, sycamore and wild cherry (Kerr 1998, Quer 1997, Oliver-Villanueva 1993, Keller 1992, Von Wedel 1964, Achterberg 1963), there is little research which links branchiness and knottiness to growth in order to create decision tools for controlling growth. Growth control is, however, an important aspect of timber quality improvement.

Despite the importance of branchiness for the production of valuable timber, in many forest enterprises there are no clear objectives describing the branchiness of a tree in quantitative terms such as the length of the branch free bole, the width of the knotty core or the maximum accepted diameter of dead or living branches (cf. Table 4.1). In addition, there is little knowledge concerning the mutual dependence of

Table 4.1. Examples of different production objectives for common ash, sycamore and wild cherry.

	harvesting diameter [cm]	rotation length [y]	length of the branch free bole [m]
common ash Thill (1975)	70	70	-
sycamore Kerr and Evans (1993)	40–60	60–70	-
wild cherry Pryor (1988)	e.g. 50	65–75	> 5, with artificial pruning

branchiness and diameter growth on site quality (cf. Thies and Hein 2000). What makes this even more surprising is that the European norm for the timber quality of common ash, sycamore and wild cherry covers numerous aspects of quality, also includes branchiness and even prescribes many margins in detail, but no indications are given on how to grow trees according to these demands (NHM 1997, also see Chapter 4.5).

From the point of view of growth control and quality, it is apparent that there are different silvicultural traditions in Europe. For instance, artificial pruning of common ash and sycamore is only occasionally practiced in German speaking countries whereas there is more experience with these techniques in France (e.g. Boulet-Gercourt 2000, Armand 1995, Duflot 1995). As for intensively managed wild cherry, branchiness is controlled exclusively by artificial pruning (cf. Spiecker 1994, Röös 1990, Spiecker and Spiecker 1988. See also textbooks on silviculture: e.g. Burschel and Huss 1997, Kerr and Evans 1993, Lanier 1986). This reflects a major difference in pruning dynamics between these three species: common ash and sycamore are broadleaves with fast natural pruning dynamics while wild cherry and poplar (*Populus* sp.) require continuous artificial pruning due to the slow shedding of dead branches on their trunks.

Thus, there is some uncertainty concerning the relation between growth and branchiness. In the following section the factors influencing branchiness are discussed in order to better illustrate how the branchiness of a tree is influenced by its environment.

Factors influencing pruning dynamics

Environmental conditions influence branching characteristics of trees. The genetic structure of the tree also affects the growth and death of the branches. Whereas environmental conditions can be controlled to a certain extent through silvicultural strategies, the genetic structure of a tree is not subject to change. Especially for wild

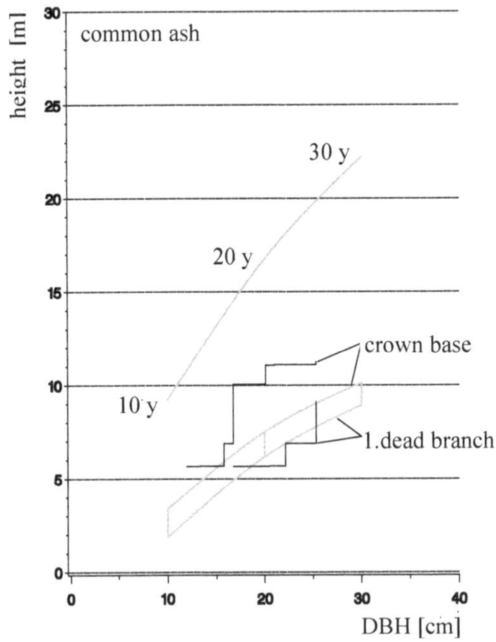

Figure 4.12. Development of the crown base and the height to the first dead branch on a tree with a mean radial increment at breast height of 4 mm/y plotted against the modeled values of crown base and length of the branch free bole at a mean radial increment of 5 mm/y (model: cf. section 2.1 of this chapter).

cherry, the selection of good planting material from the best provenance is mentioned by many authors as a very efficient measure for improving natural pruning dynamics (e.g. Armand 1995, Spiecker 1994, Röös 1990, Ducci et al. 1988, Pryor 1988).

Competition is also an eminent factor in the environment; it plays an important role in pruning dynamics mainly through shading. The growing space available for trees influences crown expansion and therefore also branch development. Thus, mixtures, planting patterns, height growth, as well as tending and thinning will affect the pruning of a tree. Because pruning dynamics and diameter growth both depend on the growing space available, both can be controlled through tending and thinning. Heavy thinning will lead to trees with a higher radial increment and a lower crown base whereas in dense stands trees will display less radial increment but a higher crown base or a longer branch free bole length (Spiecker 1991) (Figure 4.12).

Recession of the crown does not develop uniformly or continuously (Figure 4.12). If a single branch of a whorl remains alive for a relatively long time, this may result in a low crown base. With its sudden death the crown base may rise several meters to the next living branch. If annual height increment is large, due to good site conditions, the changes in crown base or the length of the lowest green branch may be significant. A mean radial increment as an average value for the whole life of a tree may also be insufficient to fully account for differences between the modeled values of the crown base or the length of the first green branch. Fluctuations in the yearly radial increment could affect a more rapid change in the dying of branches on the trunk over a certain period and may partially explain deviations from the modeled values.

In contrast to competition, site quality cannot be controlled directly through silvicultural strategies. However, a choice about height growth and thus also about the speed of dying of the lowest branches at the crown base is made by assigning different tree species to specific site conditions. For a given species the length of the branch free bole will increase with increasing site quality. Thus, natural pruning dynamics are highly site dependent.

The following section covers only natural pruning dynamics of first order primary branches attached at the trunk. Other aspects of branchiness, such as epicormics and their appearance in the different species, is not discussed here. These aspects are described in more detail by Hein (2004). For epicormics on wild cherry some remarks can be found in Pryor (1988) and Röös (1990).

4.2.2 Modeling pruning dynamics

Modeling the length of the branch free bole

As described in the previous section, changes in competition can lead to changes in the height of the crown base and the length of the branch free bole. Site conditions may also influence pruning dynamics. The following section presents an allometric model to predict pruning dynamics depending on competition and site quality.

Data for this study were collected for common ash, sycamore and wild cherry in thirteen European countries. The data cover a geographically wide area from England to Poland and from Denmark to Croatia (Hein 2004). The data cover a geographical area from 45° 30‘ to 55° 31‘ northern latitude and from -2° 57‘ western to 18° 02‘ eastern longitude, which corresponds to an extension of 1500 km from west to east and 1300 km from north to south. Only dominant, co-dominant, sub-dominant was measured. No suppressed trees were measured. The methodology followed a retrospective approach, using the principle of space-for-time substitution (i.e. chronosequences) with temporary plots of different developmental phases each with twenty-five trees. A total of 919 common ash and 632 sycamore trees with a wide range of tree dimensions were available for the analysis (Table 4.2). Since the possibility that wild cherry had been artificially pruned on many plots could not be excluded, data for this species were not included in the following calculation so as to prevent erroneous predictions of the natural pruning dynamics. The data covered measurement of tree height, height of the first living primary branch (crown base), height of the first dead primary branch, diameter at breast height and tree age. In addition, a special remark, “forked tree” was taken note of. A fork was defined as the lowest living branch that reached at least 50% of the diameter of the trunk which led to the top of the tree.

For 57% of the common ash trees (sycamore: 66%), the length of the branch free bole ended with a first green branch and no dead branches were found below. A

Table 4.2. Range in data of the variables used in the model on the length of the branch free bole and for additional variables describing the data set.

variable	lower limit ash	lower limit maple	upper limit ash	upper limit maple	arith. mean ash	arith. mean maple	standard dev. ash	standard dev. maple
length of the branch free bole: [m]	1.6	0.2	28.4	18.7	10.7	8.7	4.95	3.89
tree height: [m]	8.1	5.1	45.8	37.8	24.4	21.7	7.80	6.77
DBH: [cm]	3.5	3.8	96.5	88.8	30.9	31.1	18.09	16.73
forked tree (non-dimensional, two discrete values)	0 (= not forked)		1 (= forked)		–		–	
tree age: [y]	10	10	205	179	58	61	35.89	34.29
mean radial increment at 1.3m: [mm]	0.8	1.1	6.1	6.0	2.8	2.7	0.90	0.77

Table 4.3. Results of the stepwise regression and the variance analysis for the variables of the model on the length of the branch free bole, all parameter estimates significant at $\alpha = 0.001$*; common ash: r^2adj. = 0.76; sycamore: r^2adj. = 0.68.

variable	parameter estimate ash	parameter estimate maple	standard error of the estimate ash	standard error of the estimate maple	partial r^2 ash	partial r^2 maple
intercept	-3.763*	-1.443*	0.291	0.306	-	-
tree height: [m]	0.771*	0.634*	0.018	0.021	0.60*	0.54
DBH: [cm]	-0.172*	-0.150*	0.009	0.010	0.09*	0.12
height of the tree *forked: h_{tree}*0/1	-0.119*	-	0.009	-	0.06*	-
forked tree (non-dimensional, two discrete values)	-	-1.826*	-	0.340	-	0.01
tree age: [y]	0.024*	0.019*	0.004	0.005	0.01*	0.01

linear multiple regression model was established to predict the length of the branch free bole with tree height, diameter at breast height, tree age and a parameter (0 or 1) for a forked or non-forked tree as independent variables (Table 4.3). Using a stepwise regression 76% of the total variation of the dependent variable of common ash trees could be accounted for (sycamore: 68%). Most of the variation could be explained by the height of the tree, the diameter at breast height and, for common ash, the interaction of tree height and the occurrence of forks. Tree age contributed only very little to the reduction of the total variation and had no significant influence for sycamore. No significant interaction between the independent variables could be found besides the aforementioned interactions. As predictions for forked trees represent the specific growth condition of the temporary sample plots within the study, the model for the length of the branch free bole should provide predictions only for non-forked trees. These estimates could vary depending on the data.

The combined effect of DBH and tree age (i.e. the radial increment) could be interpreted as a measure of individual tree competition. The effect of tree height together with tree age represents site quality. The model predictions can thus be regarded as sensitive to competition and site conditions.

A visual check of the residuals did not reveal any sign of systematic bias for trees from different dominance classes or for trees with different mean radial increments. The models are limited to a mean radial increment of 2 to 5 mm and a diameter at breast height from 5 to 80 cm. Because there are only a few observations for old trees with a high radial increment, a special model restriction is set to a diameter at breast height of 60 cm if the mean radial increment is more than 4.5 mm.

Production objectives for controlling branchiness

The objectives for the length of the branch free bole can be derived from the model equation. The signs of the parameter estimates indicate that the length of the clear bole increases with increasing height or age. Trees which have reached a certain height at a certain age have a shorter branch free bole if they have a larger diameter at breast height (due to less competition). Since height and age were included in the model, an interpretation relative to site quality is possible. Table 4.4 displays production objectives with regard to the desired harvesting diameter, the mean radial increment and site quality. Natural pruning dynamics of both species are so fast that on very good sites, even at a mean radial increment of 4 mm, a length of the branch free bole of 16.0 m or 11.8 m can be reached for common ash or sycamore respectively. Because no similar model for wild cherry is available, a special remark for this species is given here: wild cherry shows a strongly deviating behavior in its natural pruning dynamics. Although its branches soon die when shaded, they are retained for many years after their death even if the diameter of the branch is small. Therefore, most guidelines for pruning control recommend artificial pruning (e.g.: MELF-BW 1999, Allegrini et al. 1998, Armand 1995, Spiecker 1994, Röös 1990, Pryor 1988).

Because a higher absolute length of the branch free bole can be expected on good sites, possible values for the length are given as relative, and thus site dependent, values. Guidelines for artificial pruning of wild cherry from the point of view of growth control are given in 4.2.3.

The main growth parameters for common ash and sycamore are shown in Figure 4.13 to better present the dimensions of a tree that has reached such production objectives. It should be noted that even for a mean radial increment of 3 mm the

Table 4.4. Production objectives (length of the branch free bole) for three different site qualities for common ash, sycamore, and wild cherry. Site index: common ash: 33, 27, 21 m at age 60 yrs; sycamore: 30, 24, 18 m at age 60 yrs (Hein 2004) and wild cherry: 27.5, 24.5, 21.5 m at 50 yrs (Röös 1990). Crown cover: 70%.

	harvesting diameter [cm]	mean radial increment [mm/y]	length of the branch free bole [m]		
			site index: 33 m	site index: 27 m	site index: 21 m
common ash		2	-	-	-
		3	20.3	15.7	10.8
	60	4	16.0	11.3	6.6
		5	12.8	8.2	-
			site index: 30 m	site index: 24 m	site index: 18 m
sycamore		2	-	-	-
		3	14.4	10.9	7.1
	60	4	11.8	8.0	-
		5	9.7	5.9	-
			site index: 27.5 m	site index: 24.5 m	site index: 21.5 m
wild cherry	60	2–5		20–50%	

relative crown length does not fall below 50%. This could be seen as an indicator that long crowns are needed to maintain a high level of diameter growth. Trees of different mean radial increments exhibit different crown widths and consequently a variation in the number of crop trees with the desired diameter can be found on any one hectare. Thus, the numbers of crop trees per hectare are adopted from Chapter 4.3.

Pruning dynamics of open grown trees

Trees situated in free growing conditions reveal the growth potential of a species under specific site conditions where there is no competition from neighbouring trees. With respect to natural pruning dynamics these trees show the minimum length of the branch free bole or height of the crown base a tree can obtain under given environmental circumstances. They can thus be used as a reference or a growth limit for natural pruning. In addition, open grown trees can help answer the question as to whether crown recession can be stopped completely through heavy thinning. For this analysis, data for 42 open grown common ash and 32 open grown sycamore trees were collected in several European countries. To ensure that the

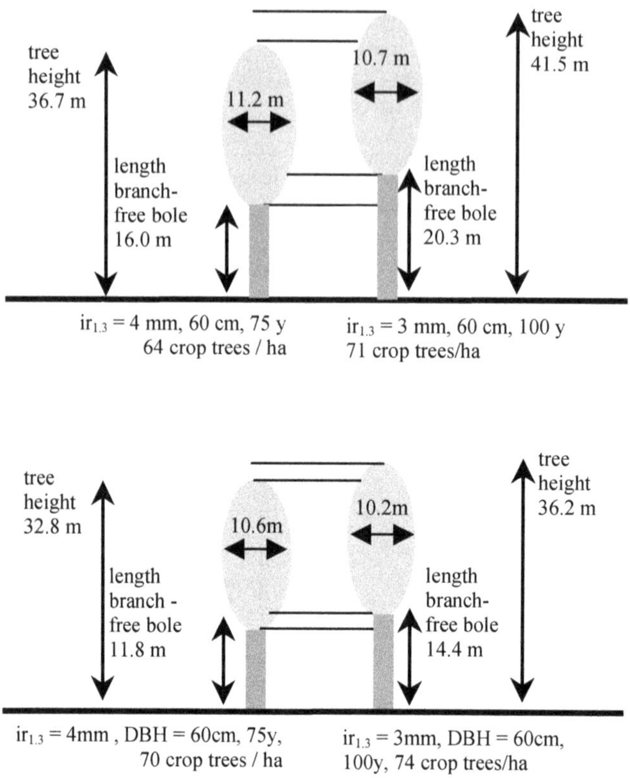

Figure 4.13. Target dimensions for common ash (above), site index: 33m, dominant height at 60y. Sycamore (below), site index: 30 m. The figures for crown width and number of trees/ha are taken from Chapter 4.3 in this book.

trees chosen had grown predominantly free from competition throughout their lives, trees with a first green branch higher than 5 m on any side were excluded. For this reason one-sided suppressed trees were also excluded. No open grown wild cherry trees were available. A more detailed description of the data is given in Table 4.5.

In contrast to trees from stands, only a few of the open grown trees had dead branches below the crown base: 4 of the 42 open grown common ash trees and 1 of 32 sycamore. On average these dead branches were 0.3 m below the first living branch for ash and 0.7 m for sycamore. Because these differences were small, no separate model for the crown base and the length of the branch free bole was set up. The model presented in the following paragraph represents the model of the first living primary branch, and this is thus the crown base model. The crown base of open grown trees covers a range from 0.4 to 4.9 m for common ash and 0.2 to 3.1 m for sycamore. The clear bole length model for stand grown trees did not correctly predict the height of the crown base of open grown trees. A untransformed linear regression fitted the data best (Table 4.6). The height to the crown base increases slightly with increasing diameter at breast height (Figure 4.14).

Table 4.5. Range in data of common ash and sycamore trees for the variables used in the model on crown base for open grown trees and for additional variables describing the data set.

	lower limit		upper limit		arith. mean		standard dev.	
variable	ash	maple	ash	maple	ash	maple	ash	maple
height of the crown base: [m]	0.4	0.2	4.9	3.1	2.1	1.5	0.89	0.82
DBH: [cm]	3.5	2.0	102.5	132.0	40.9	43.4	22.75	29.04
tree height: [m]	3.4	2.3	30.4	23.0	13.5	11.7	5.15	5.20
tree age: [y]	7	5	169	162	49	58	37.06	38.20
mean radial increment: [mm]	1.48	1.85	8.27	6.5	4.67	3.62	1.62	1.36

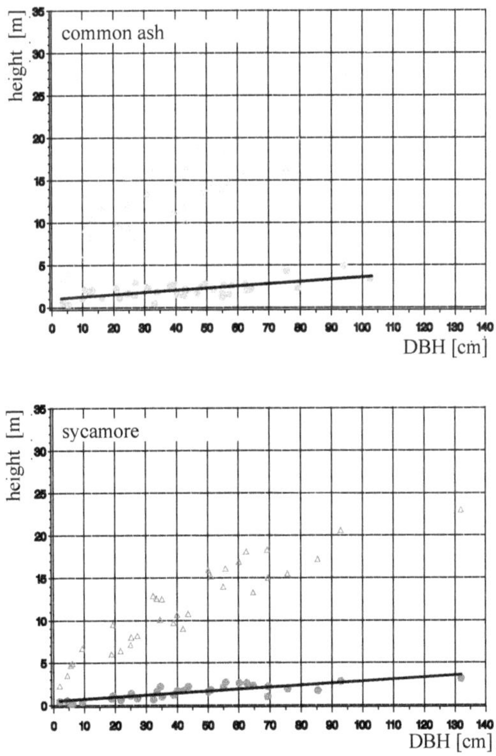

Figure 4.14. Tree height and height of the crown base of common ash and sycamore open grown trees; parameters of the regression line: cf. Table 4.6.

Table 4.6. Results of the linear regression for common ash and sycamore for the variables of the crown base model for open grown trees, *= significant at $\alpha = 0.0001$, ** = significant at $\alpha = 0.005$; crown base = f(DBH).

	parameter estimate		standard error of the estimate		r^2	
variable	ash	sycamore	ash	sycamore	ash	sycamore
intercept	1.012*	0.512**	0.216	0.155	-	-
DBH: [cm]	0.026*	0.023*	0.005	0.003	0.43* adj.: 0.42	0.66* adj.: 0.65

Table 4.7. Modeled height of the crown base and relative crown length of open grown common ash und sycamore trees with a diameter at breast height of 60 cm; no wild cherry open grown trees were available.

	common ash	sycamore
DBH: [cm]	60	60
Height of crown base: [m]	2.6	1.9
Tree height: [m]	28	26
Relative crown length: [%]	90.7	92.7

This effect could be interpreted as a result of self-pruning through mutual shading of the branches themselves, with the result that branches at the crown base die off. However, it could not be excluded that agricultural activities close to the open grown tree or the browsing by animals lead to the loss of branches in early stages of tree development. In addition, the occurrence of dead branches was clearly linked to the shading by hays around the open grown tree. Thus, there is no clear evidence, for tree species with very good natural pruning such as common ash and sycamore, that losing primary branches is an effect of self-shading.

The predicted crown base for an open grown common ash tree with a diameter at breast height of 10 cm is 1.3 m and rises to 2.6 m at 60 cm (sycamore, 10 cm: 0.7; 60 cm: 1.9 m). This is very close to the predictions of a similar model by Hasenauer (1997) for common ash and sycamore based in Austria based on a smaller sample size. Assuming a mean radial increment of 7 mm for both species in the case of free growth, a diameter at breast height of 60 cm will be reached in 43 years (cf. Hein 2004). Ash trees with a height of 28 m at that age will have a relative crown length of 90.7%, and sycamore open grown trees a somewhat longer relative crown length: 92.7% (Table 4.7).

4.2.3 Development of branches and knots

Development of the length of the branch free bole

From the model equations presented in 4.2.2 and through assumptions on stem form, the following simple rules for predicting branchiness can be derived (cf. Figure 4.15). Although no similar model for wild cherry is available, the validity of these rules for this species can be assumed:

Site quality: For a fixed radial increment, the better the site quality the longer the branch free bole. With the radial increment being equal, a better site quality induces more radical changes in the length of the branch free bole, especially at young age. For a fixed radial increment, the better the site quality, the less amount of time it takes to reach a desired absolute length of the branch free bole.

Diameter growth: For a fixed site quality, the greater the radial increment the shorter the branch free bole. For a given site quality, the greater the radial increment, the longer it takes to reach a desired absolute length of the branch free bole.

Forks: Due to their forks, forked trees have, relative to non-forked trees, a significantly shorter branch free bole.

Development of the width of the knotty core

A clear bole with no dead or living branches visible from the outside can have a large knotty core. From the perspective of wood quality, a 1/3 proportion of knotty core relative to the stem diameter at a given height can be given as a maximum requirement for high quality timber (Schulz 1959).

Using species specific taper functions (Dagnelie et al. (1999) for *F. excelsior* L., In et al. (1972) for *Prunus avium* L. and *Acer pseudoplatanus* L.) the width of the knotty core was simulated (Figure 4.16). Simple rules for predicting the internal branchiness can be derived from the model equations presented in 4.2.2 and the aforementioned assumptions:

Site quality: For a fixed radial increment and at a fixed height of the trunk, a better site quality leads to a smaller width of the knotty core. Radial increment being equal, the better the site quality, the lower the proportion of the branch free bole that is made up of a knotty core that exceeds 1/3 of the stem diameter.

Diameter growth: For a fixed site quality and a fixed height of the trunk, the higher the radial increment the larger the width of the knotty core. Site quality being equal, the higher the radial increment, the higher the proportion of the branch free bole that is made up of a knotty core that exceeds 1/3 of the stem diameter.

As natural pruning of wild cherry is slow (Burschel and Huss 1997), most authors recommend artificial pruning (cf. references in 4.2.3). The following list gives a summary of rules for artificial pruning in order to produce high-quality timber, referring mainly to recommendations given in Allegrini et al. (1998), Boulet-Gercourt (2000) and Spiecker (1994).

Pruning rules:

- Pruning has to be done at the branch collar without injuring the wood of the trunk. Otherwise, fungi could enter into the trunk and initiate coloration.

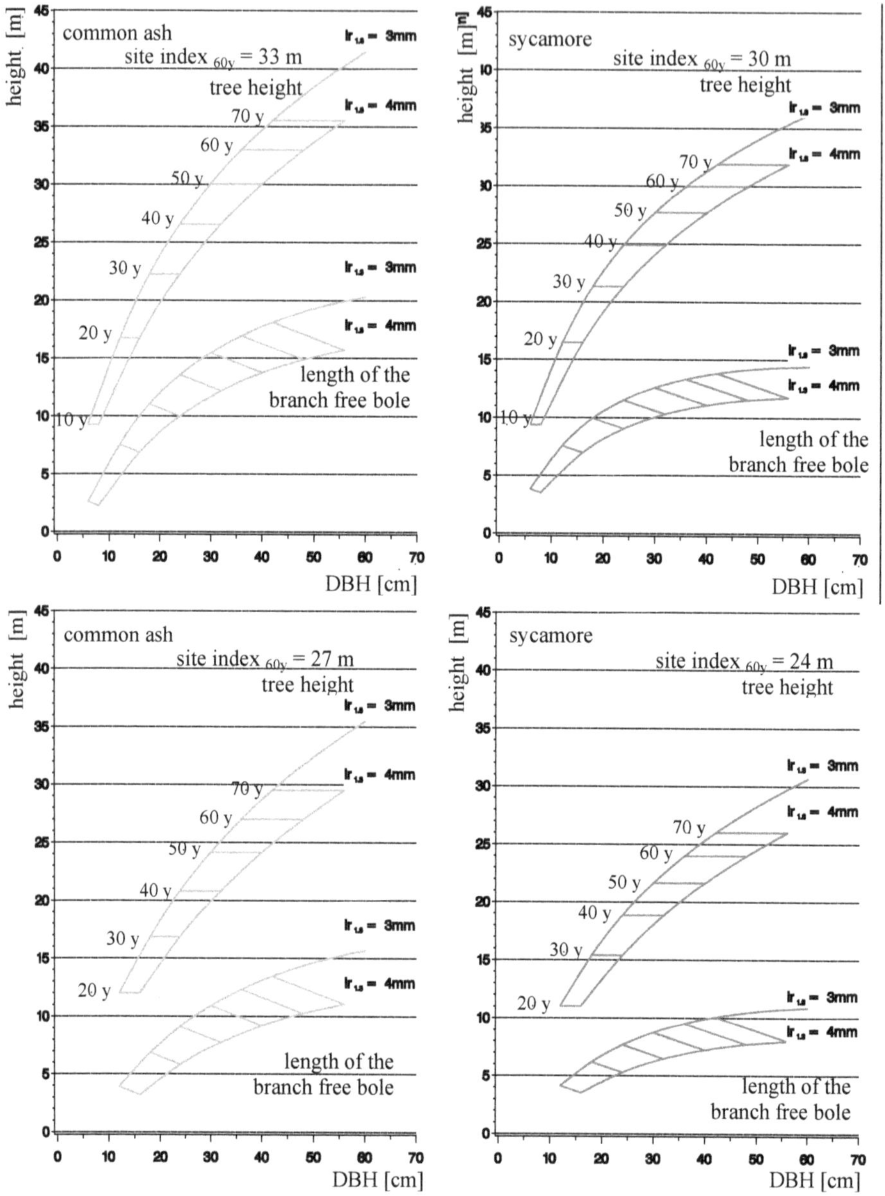

Figure 4.15. Length of the branch free bole as a function of diameter at breast height of a site index 33 m and 27 m (dominant height at 60 y, common ash, left side) and for sycamore of a site index 30 m and 24 m (dominant height at 60 y, sycamore, right side) and for scenarios with mean radial increments of 3 mm and 4 mm.

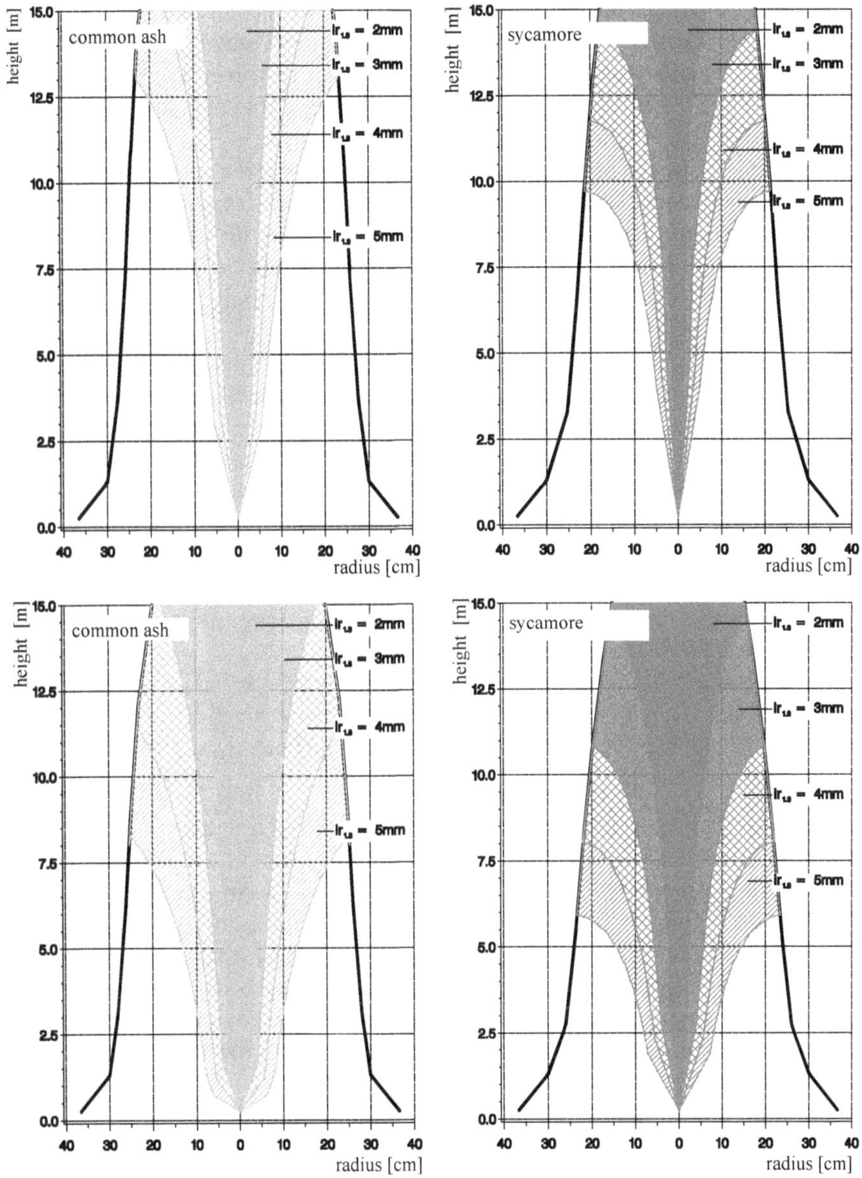

Figure 4.16. Width of the knotty core inside a trunk with a diameter at breast height of 60 cm for common ash (dominant height at 60 y = 33 m above, and 27 m below, left side) and for sycamore (dominant height at 60 y = 30 m above, and 24 m below, right side). Each is shown in a longitudinal section.

- Pruning should be done before branches reach a diameter of more than 3 cm at the collar. The tree will heal better and more rapidly without requiring additional healing products and the risk of infection with in turn be reduced. It is recommended to cut in advance those branches within the crown that will reach this diameter limit before the next pruning.
- Pruning is a frequently recurring task. Due to biological and technical reasons, it is recommended to repeat pruning every second year to prevent branches from reaching large diameters. In addition, a single abrupt and overly heavy reduction of the green crown could lead to the development of epicormics. Therefore regular and moderate interventions are better than infrequent and severe interventions.
- The number of trees to be pruned should be limited to the crop trees that will reach a final diameter. To reduce the risk of losses through self-thinning a higher number may be pruned, but only to a dominant height of 4 to 5 m.
- Wild cherry has to be pruned by cutting off dead and green branches. Pruning has to be started early in tree development to prevent thick branches (from 4 years on).
- Good results have been obtained with pruning at the end of winter, at early growth stages after final spring frost, or in summer at full growth. Pruning should not be carried out during periods of bud flushing and sap flow (from late August until the period of leaf abscission).
- Scissors are sufficient as professional tools for cutting branches with a diameter of less than 2 cm. For larger diameters a saw is convenient. Saws mounted on a telescopic pole are recommended for higher branches. Always disinfect equipment before use.

Special attention to the diameter development of the branches is required. As diameter growth of wild cherry branches is very quick, especially for those experiencing uninhibited growth, a diameter limit of 3 cm can be reached very quickly. Most branches with larger diameters exhibit dark colored heart wood that no longer prevents fungi from entering the tree. Therefore it is imperative to prune early and frequently before large branch diameters are reached.

Complementary pruning can also be conducted and prevents the use of a single criterion, such as ‘well pruned stems’, when choosing crop trees. Therefore the most robust and straight stems can be selected even if they are not well pruned. Work is conducted from ground level to heights of 6–7 m maximum.

A schematic impression of the consequences of artificial pruning on knottiness of wild cherry is shown in Figure 4.17. Frequent pruning of branches with a diameter of more than 3 cm will result in a thin knotty core up to the desired height of the branch free bole. This will result in a heavy crown release conducive to big branches above that height, where exclusively knotty wood is produced. A phase of diameter growth suppression, if carried out properly, minimizes artificial pruning which can often be costly. It occurs over a decade on trees of heights 8–15 m to allow the canopy cover to close so that lower branches will become deprived of light resources and will fall.

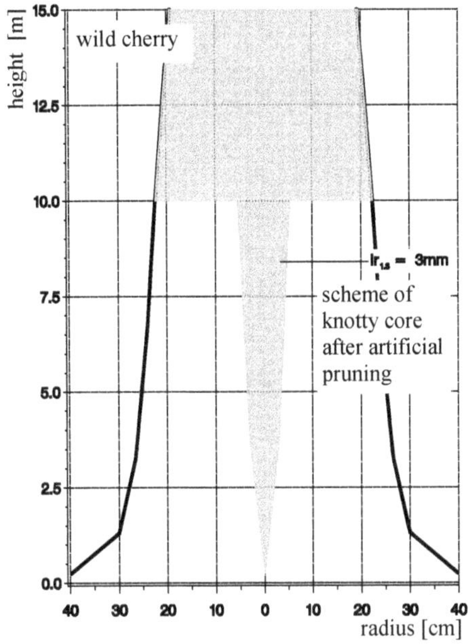

Figure 4.17. Width of the knotty core inside a trunk after artificial pruning with a diameter at breast height of 60 cm for wild cherry; shown in a longitudinal section.

A two phase concept for controlling knots

In addition to the scenarios with a fixed mean radial increment over the whole production time presented in the previous sections, a more flexible strategy can be used to control diameter growth. In a two phase system of growth control (cf. Wilhelm et al. 1999a, Wilhelm et al. 1999b, Wilhelm et al. 1999c, Wilhelm & Raffel 1993, Spiecker 1991), the forest planner, once he has made a decision regarding the length of the clear bole, final diameter of the crop tree and production time, follows the first phase of growth control.

From stand establishment to the time at which the desired length of the clear bole is reached, forest operations focus on natural pruning. During the first phase a high radial increment is not consistent with a reduction of the width of the knotty core inside the trunk and therefore is not the focus. Figure 4.18 shows the pruning dynamics of common ash on good site conditions with a mean radial increment of 3 mm. At the age of 27 years a clear bole length of 10 m is reached. The second phase concentrates all forest operations on speeding up diameter growth until the final harvesting diameter is reached. As no further crown recession is desired, the crop trees are subject to heavy selective thinnings, resulting in convergence of the radial increment to a site-depending maximal value (here: 7 mm, cf. Hein 2004). A significant reduction of the knotty core below the crown can be obtained with this scenario through a lower radial increment in earlier phases of tree development and

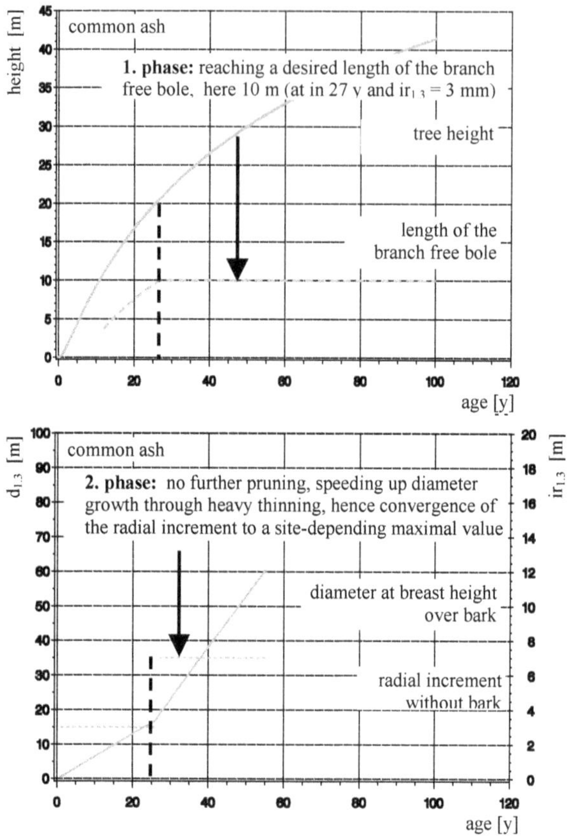

Figure 4.18. A two phase growth control concept exemplified with common ash: 1st phase: reaching a desired length of the branch free bole; here: 10 m, mean radial increment: 3mm, site index: 33 m dominant height at 60 yrs). 2nd phase: no further pruning, speeding up of diameter growth through heavy thinning and hence convergence of the radial increment to a site-dependent maximum value (here: 7 mm).

a nearly free growth after a certain length of clear bole is reached (Figure 4.19). This corresponds to a width of the knotty core as described for artificially pruned wild cherry. Special attention should be given to the transition between the first and the second phase. Like common ash, sycamore and wild cherry can form epicormics after sudden crown release (Hein 2004, Röös 1990, Pryor 1988, Otto 1987). A transition thinning is therefore recommended to prepare trees for open growth. This may also prevent trees from losing vitality as a consequence of an abrupt change in their environmental conditions.

The selection of crop trees occurs at the point of transition to the second phase when the desired length of the branch free bole is reached. As natural pruning depends on height growth (see Chapter 4.2.2), this point of development differs according to site conditions. The stage of crop tree selection is reached early in the production time for common ash, sycamore and wild cherry due to their fast height growth in youth (e.g.

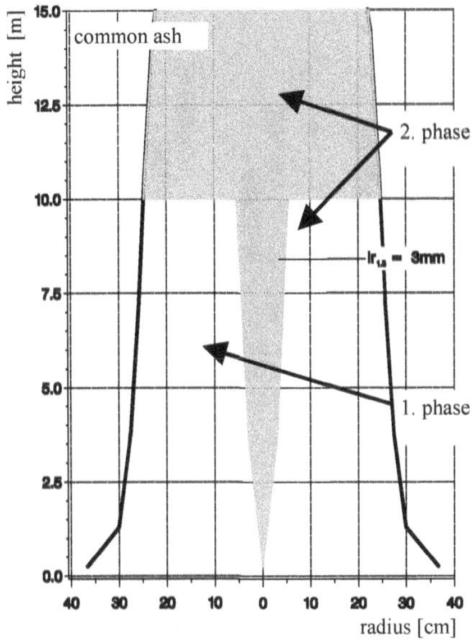

Figure 4.19. Width of the knotty core in a two phase growth control concept (site index: 33 m dominant height at 60yrs, common ash). 1st phase: natural pruning, mean radial increment of 3 mm until a desired branch free bole length is reached, here: 10 m. 2nd phase: no further pruning, speeding up of diameter growth to the site-dependent value of open grown trees and hence abandonment of natural pruning.

Röös 1990, Kjölby et al. 1958, Volquardts 1958). It is thus even more important to benefit from the high potential in diameter growth at younger ages for these trees. Many authors recommend tolerating a short branch free bole combined with artificial pruning to favor an early promotion of diameter growth and a reduction in production time. Other advantages of a short production time include a lower risk of rot which often occurs in wild cherry from 50 years on (e.g. Spiecker 1994, Röös 1990) or the occurrence of black heart in common ash (e.g. Kerr 1998, Thill 1970).

4.2.4 Conclusions

The models of branching dynamics presented here allow one to better set quantitative management objectives for wood quality. The main variables are the length of the branch free bole and the width of the knotty core. Another central decision a forest enterprise has to make, and that is closely linked to branching, is that of the target diameter, production time and hence stem radial growth. As this refers to the number of crop trees per hectare (cf. Chapter 4.3), some conclusions have to be drawn concerning the production of externally branch free wood volume on a hectare basis. This could also highlight species-specific differences in pruning dynamics, crown width development and stem form.

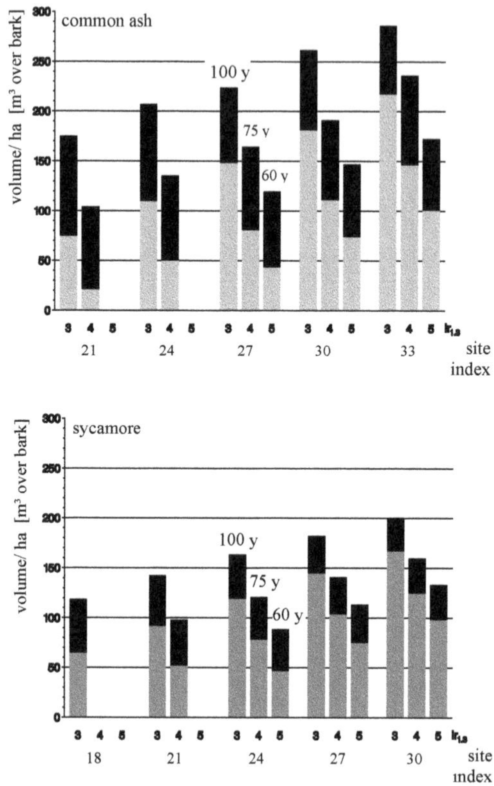

Figure 4.20. Volume of branch free wood per hectare over the whole length of the externally branch free bole of the crop trees with a target diameter of 60 cm at breast height (scenarios of mean radial increment from 2 to 5 mm, site index according to Hein (2004)). Symbols: black: volume of externally branch free wood of the crop trees with a width of the knotty core exceeding 1/3 of the diameter of the trunk. Crown cover: 70%. No values are given for scenarios with mean radial increments outside the scope of application of the models.

The model presented in this chapter and the crown width model in Chapter 4.3 can quantify these relationships (Figure 4.20; both species). The length of the branch free bole decreases with an increasing mean radial increment, and there will be a smaller number of crop trees per hectare with a target diameter at breast height of, for instance, 60 cm, than if a lower target diameter was selected. The volume of externally branch free wood on a hectare basis decreases as a result. For a given mean radial increment the volume of externally branch free wood per hectare increases with increasing site quality.

A comparison of both species shows that the productivity of common ash, as far as the per hectare volume of externally branch free wood is concerned, is higher than that of sycamore. This reflects differences in pruning dynamics, number of crop trees per hectare and stem form. Two other trends also become apparent when considering in more detail the volume-related proportion of the externally branch free wood with a knotty core exceeding 1/3 of the diameter of the trunk: firstly, with an increasing mean radial increment the portion of this volume increases as well; and secondly, with increasing site quality the portion of the externally branch free wood with a knotty core exceeding 1/3 of the diameter of the trunk decreases at a fixed mean radial increment. The per hectare productivity of externally branch free common ash wood is higher than that of sycamore, all else being equal. However, the proportion of the externally branch free wood with a knotty core exceeding 1/3 of the diameter of the trunk is higher in the case of common ash.

References

Achterberg, W. 1963. Criteria of Roundwood Assortment and Veneer Quality with Maple, Elm and Birch. Ph.D., Universität Berlin. (Original in German).

Allegrini, C.H., Boistot-Paillard, R., Bouvet, J.Y., Depierre, A. and Mourey, J.M. 1998. The Valuable Broadleaves in Franche-Comté, Société Forestière de Franche-Comté (eds.), Besançon. (Original in French).

Armand, G. 1995. Valuable Broadleaves – Silviculture of Plantations on Forest Sites – Wild Cherry, Sycamore, Ash and American Red Oak. Editions IDF, Institut pour le Developpement Forestier (eds.) Paris. (Original in French).

Boulet-Gercourt, B. 2000. Le Merisier. Editions IDF, Institut pour le Developpement Forestier (eds.), Paris. (Original in French).

Burschel, P. and Huss, J. 1997. Principles of Silviculture: a Guideline for Studying and Practice. Parey-Verlag, 2nd edition, Berlin (Original in German).

Dagnelie, P., Palm, R., Rondeux, J. and Thill, A. 1999. Volume Tables of Trees and Forest Stands. Presses Agronomiques de Gembloux, 2nd edition, Gembloux. (Original in French).

Ducci, F., Tocci, A. and Veracini, A. 1988. Synthesis of the Inventory on *Prunus avium* L. in Northern Central Italy, Basilicata e Calabria. Annali dell' Instituto Sperimentale per la Selvicoltura, Arezzo Vol. XIX: 266–303. (Original in Italian)

Duflot, H. 1995. Liberty for Ash. Editions IDF, Institut pour le Developpement Forestier (eds.), Paris. (Original in French)

Hasenauer, H. 1997. Dimensional Relationships of Open-Grown Trees in Austria. Forest Ecology and Management 96: 197–206.

Hein, S. 2004. Controlling Natural Pruning and Diameter Growth with Common Ash (*Fraxinus excelsior* L.) and Sycamore (*Acer pseudoplatanus* L.). Ph.D. Universität Freiburg. (Original in German)

In, K., Roneux, J. and Thill, A. 1972. Dendrometrical Studies on Sycamore (*Acer pseudoplatanus* L.) and Wild Cherry (*Prunus avium* L.). Bulletin des Recherches Agronomiques de Gembloux 7: 134–190 S. (Original in French)

Keller, R. 1992. The Wood of Maple Species: State of the Art, Factors of Variability and Technological Properties. Revue Forestière Française XLIV: 133–141. (Original in French)

Kerr, G. 1998. A Review of Black Heart of Ash (*Fraxinus excelsior* L.). Forestry 71: 49-56.

Kerr, G. and Evans, J. 1993. Growing Broadleaves for Timber. Forestry Commission Handbook, 9, London.

Kjölby, V., Sabroe, A.S. and Moltesen, P. 1958. ÆR (*Acer pseudoplatanus* L.). Dansk Skovforening, Kopenhagen. (Original in Danish)

Lanier, L. 1986. Principles of Silviculture. École Nationale du Génie Rural des Eaux et des Forêts, Nancy. (Original in French)

MELF-BW. 1999. Guidelines of State-Wide Forest Types. Ministerium für Ländlichen Raum, Ernährung, Landwirtschaft und Forsten, Landesforstverwaltung Baden-Württemberg (eds.), Stuttgart. (Original in German)

NHM. 1997. German Norm: Broadleaved Roundwood, Quality-Assortment, Part 3: Ash and Maple, German Version of EN 1316-3: 1997. NHM (Normenausschuss Holzwirtschaft und Möbel im Deutsches Institut für Normung, DIN) (eds.). (Original in German)

Oliver-Villanueva, J. V. 1993. Wood Properties of Ash (*Fraxinus excelsior* L.) and its Variability with Regard to Age and Growing Space. Ph.D., Universität Göttingen. (Original in German)

Otto, H. 1987. On the Silviculture of Wild Cherry. Forst- und Holzwirt 42: 44-45. (Original in German)

Pryor, S.N. 1988. The Silviculture and Yield of Wild Cherry. Forestry Commission Bulletin 75, London.

Quer, M. 1997. On the Influence of Silvicultural Treatment on the Use of Ash-Sawn Timber (*Fraxinus excelsior* L.). Ph.D., Universität Göttingen. (Original in German)

Röös, M. 1990. The Growth of Wild Cherry (*Prunus avium* L.) in Nordrhine-Westfalia and Neighbouring Regions. Ph.D., Forstwissenschaftlicher Fachbereich Universität Göttingen. (Original in German)

Schulz, H. 1959. Untersuchungen über die Bewertung und Gütemerkmale des Eichenholzes aus verschiedenen Wuchsgebieten. Schriftenreihe der Forstlichen Fakultät der Universität Göttingen und Mitteilungen der Niedersächsischen Forstlichen Versuchsanstalt n°23, Sauerländer, Frankfurt a.M. (Original in German).

Spiecker, H. 1991. On the Controlling of Diameter Growth and Natural Pruning of Sessile and Pedunculate Oak (*Quercus petrea* (Matt.) Liebl. and *Quercus robur* L.). Schriftenreihe der Landesforstverwaltung Baden-Württemberg 72, Stuttgart. (Original in German)

Spiecker, M. 1994. Growth and Silviculture of Valuable Wild Cherry Trees. Mitteilungen der Forstlichen Versuchs- und Forschungsanstalt Baden-Württemberg 181. (Original in German)

Spiecker, M. and Spiecker, H. 1988. Silviculture of Wild Cherry Valuable Timber. Allgemeine Forstzeitschrift 20: 562–566. (Original in German)

Thies, M. and Hein, S. 2000. Questionnaire of Experts on the Role of Valuable Broadleaved Species in German Speaking Countries. Sektion Ertragskunde im Deutschen Verband Forstlicher Forschungsanstalten, Jahrestagung 2000, Kaiserslautern. Pp. 256–269. (Original in German)

Thill, A. 1970. Common Ash and its Silviculture. Gembloux, Éditions Duculot S.A., Les Presses Agronomiques de Gembloux A.S.B.L, Mémoires 8. (Original in French)

Thill, A. 1975. Contributions to Common Ash, Sycamore and Wild Cherry (*Fraxinus excelsior* L., *Acer pseudoplatanus* L., *Prunus avium* L.). Bulletin de la Societé Royale Forestière de Belgique 82: 1–12. (Original in French)

Volquardts, G. 1958. Common Ash in Schleswig-Holstein in Germany. Ph.D., Universität Göttingen Hann. Münden. (Original in German)

Von Wedel, K. 1964. Investigations on Wood Properties and Use of Maple Wood. Ph.D., Universität Göttingen Hann. Münden. (Original in German)

Wilhelm, G.J., Letter, H.A. and Eder, W. 1999a. The Phase of Controlling Diameter Growth. Allgemeine Forstzeitschrift/ Der Wald, 5/1999: 236–238. (Original in German)

Wilhelm, G.J., Letter, H.A and Eder, W. 1999b. Growth Objectives and Silvicultural Principles. Allgemeine Forstzeitschrift/ Der Wald, 5/1999: 232–233. (Original in German)

Wilhelm, G.J., Letter, H.A. and Eder. W. 1999c. The Phase of Maturing. Allgemeine Forstzeitschrift/ Der Wald, 5/1999: 239–240. (Original in German)

Wilhelm, G.J. and Raffel, D. 1993. Silviculture of Time-Limited Mixtures of Wild Cherry and Beech on the Lorrain Plateau. Revue Forestière Francaise, Band XLV–6: 651–668. (Original in French)

4.3 Controlling Diameter Growth of Common Ash, Sycamore and Wild Cherry

Sebastian Hein and Heinrich Spiecker
Institute for Forest Growth, University of Freiburg, Germany

Abstract

Quality and dimension determine the value of a log. In this chapter an allometric model on crown width for common ash (*Fraxinus excelsior* L.), sycamore (*Acer pseudoplatanus* L.,) and wild cherry (*Prunus avium* L.) is used to set up decision tools for controlling diameter growth. The model is used to show the link between production time, the number of mature trees at the end of production time and their interactions with diameter at breast height. With a separate analysis of crown width for open grown trees the maximum crown expansion for a given diameter at breast height can be calculated, and the number of released crop trees per hectare at the end of the production time can be derived. Based on the crown width model several tools for selective thinning with crop trees are developed.

Keywords: crop trees, diameter growth, open grown trees, decision tools, Fraxinus excelsior L., Acer pseudoplatanus L., Prunus avium L.

4.3.1 Producing valuable timber

Does diameter matter?

Prices on the wood market for broadleaves are to a large extent determined by two main parameters: the quality of the logs and their dimension. This also applies to valuable logs of broadleaves such as common ash (*Fraxinus excelsior* L.), sycamore (*Acer pseuodplatanus* L.) and wild cherry (*Prunus avium* L.). Changes in demand and supply caused by the general economic conditions, natural hazards, changing fashions, mode of sales etc. have an impact on the market value of logs as well. However, the quality and dimension of the logs are decisive parameters that can be controlled by forest management.

Results of several surveys in the UK, France, and Germany clearly show an increase in price per cubic metre with increasing tree diameter (Soulères 1997, Spiecker 2001, 2003, Whiteman et al. 1991) for broadleaves in general, but also for the three aforementioned tree species. The increase in price per unit of volume is highest for small sized trees and decreases with increasing tree size (Whiteman et al. 1991). From these observations an initial conclusion concerning the importance of the size of the log brought to the market can be drawn: production of timber yielding good prices on the wood market must focus on the production of large sized trees. The log size finally targeted depends on the individual preferences of the forest owner, the growing conditions and the species specific age related risks.

The implications of diameter growth on quality and especially on natural pruning dynamics must not be neglected. An accelerated diameter growth may result in insufficient quality of the crop: epicormics after sudden heavy release, thick branches, and a large knotty core (Spiecker 1991). When formerly released trees experience strong inter-tree competition large branches may die. These branches reduce wood quality and may be the gateway for fungi causing wood decay. An accelerated diameter growth may not always result in undesirable qualities however. In the case of common ash, large growth rings even improve the mechanical properties of the wood (cf. Chapter 4.5).

To better reflect the current variety of targeted harvesting diameters of mature trees, the following section discusses diameter objectives as referred to in the literature. Based on these objectives methods for forest management with special emphasis on diameter growth control are derived. A short description of the most important factors influencing diameter growth is given.

Diameter objectives for common ash, sycamore and wild cherry in Europe

To assess the targeted diameter objectives of mature trees before harvesting, two data sources were used: an international survey on the management of valuable broadleaved forests as presented in Chapter 2.2 and a literature review. The harvesting diameters of mature trees targeted by private and public forest owners reflect their wide range of individual preferences (cf. Chapter 2.2). Indeed, the astonishing differences between the individual objectives may partly be explained by a lack of knowledge and experience concerning the potential of the species for diameter development.

Production objectives in Europe differ significantly between the authors. Table 4.8 shows some examples. Besides a different perception of the growth potential of the three species, differences in the silvicultural traditions between the countries may also be apparent (cf. Barret and Holmsgaard 1964).

Because diameter at breast height is linked to age through the yearly diameter increment, an indication of production time can be given. In some literature the production time is not clearly described. As diameter growth is linked to crown width expansion, some references also indicate the number of crop trees per hectare expected at the end of rotation time. But only a few authors give quantitative information on the targeted length of the clear bole.

Thus there is a significant degree of uncertainty concerning the relation between harvesting diameter, production time and other elements of the production objectives. To quantify this relation the following section discusses the factors influencing diameter growth. A focus is placed on factors that can be used to control diameter growth in order to produce valuable timber.

Table 4.8. Examples of different production objectives for common ash, sycamore and wild cherry.

	harvesting diameter [cm]	production time [y]	number of crop trees/ ha [-]	length of branch free bole [m]
common ash Pilard-Landeau and Le Goff (1996)	60–70	60	60	8 (h_0 = 30 m)
sycamore Kerr and Evans (1993)	40–60	60–70	140–170	-
wild cherry Armand (1995)	> 50	50–60	70–90	5–6 (with pruning)

Factors influencing diameter growth

A tree's current diameter growth reflects the growing conditions. Important growing conditions are the site and the growing space. These conditions may change over time. In addition the growth may be influenced by the genetic structure of the tree. Short term variability of climatic conditions causes changes in annual diameter increment. Long-term changes in site productivity may have an impact especially on the height increment. While these site factors are generally not controlled, competition offers the opportunity to control diameter growth according to selected objectives.

The removal of trees around a selected crop tree reduces inter-tree competition and increases the growing space available for the subject tree. As a consequence its diameter growth is enhanced. The subject tree retains its dominant position within the stand or can even become dominant over its neighbors, if not already so. Thus, differences in radial increment at breast height reflect the "social status" of the tree. Figure 4.21 illustrates the level of radial increment of two sycamore trees measured on stem disks at breast height. Over the whole observation period the dominant trees showed a higher radial increment than the sub-dominant tree due to their larger available growing space.

The significant effects of growing space on diameter growth can be observed very early after the establishment of the stand (Figure 4.22). Five years after stand establishment, the radial increment of a dominant common ash tree from a stand with a wide initial spacing surpasses the radial increment of an ash tree from a dense initial spacing. Continuous thinning can ensure a high level of diameter growth over a long period of time.

Trees with larger crowns grow faster and tend to maintain a faster growth rate as compared to trees with smaller crowns even at an old age. Released wild cherry trees, which grew fast during an early period, will most likely continue to grow at a higher rate during later periods (Figure 4.23). Although crown release generally accelerates

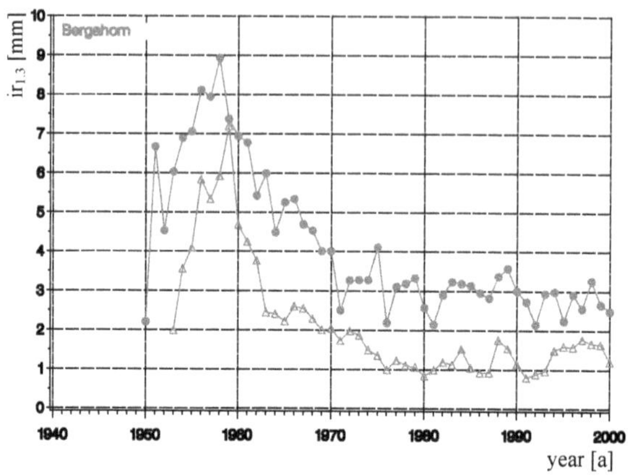

Figure 4.21. Annual radial increment at breast height ($ir_{1.3}$; stem disks taken at 1.3 m) of the mean of two dominant and of one sub-dominant sycamore tree over time. In these cases the increment decreases when the trees get older. The increment level of the dominant trees is considerably higher. Symbols: dot: dominant tree; triangle: sub-dominant.

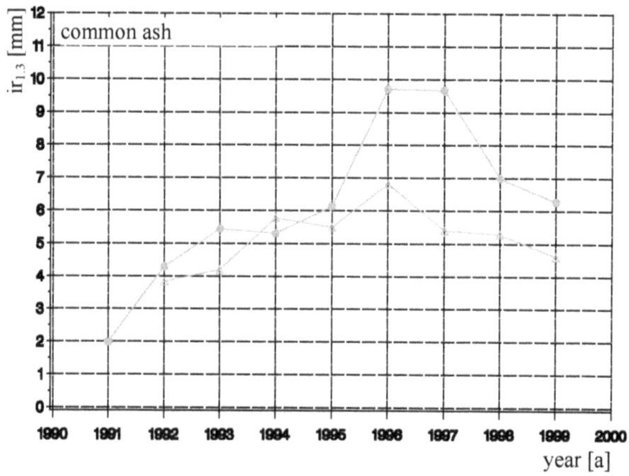

Figure 4.22. Annual radial increment at breast height ($ir_{1.3}$) of two dominant common ash trees from stands of different initial spacing. Wider spacing accelerated radial increment. Symbols: triangle: initial spacing (1990) 4.5×4.0 m common ash and 2.25×2.0 m black alder; dot: initial spacing (1990) 4.5×4.0 m pure common ash.

growth of trees of all sizes, trees with larger crowns continue to grow faster after release than trees with smaller crowns that have been released. The periodical diameter increments of 13 wild cherry trees of up to 80 years of age indicate that a mean annual diameter increment of more than 6 mm can be reached. The extent to which old wild cherry trees are capable of responding to crown release is still open to question, although some authors indicate that the potential diameter growth decreases significantly with age (e.g. Kerr and Evans 1993, Pryor 1988).

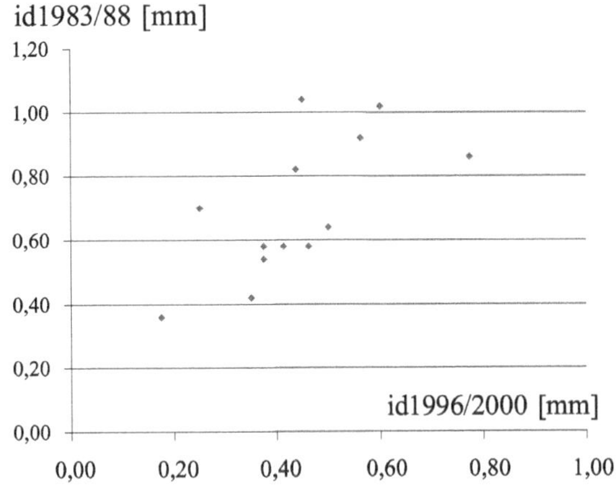

Figure 4.23. The periodical annual diameter increment at breast height (id) of wild cherry trees with released crowns in two periods 1983/88 and 1996/2000 is compared. The trees were 80 years old at the end of 2000 (further information see Spiecker 1984). Trees tend to maintain their growth rate.

4.3.2 Production goals and controlling diameter growth

The crown width – diameter at breast height relationship

As described in the previous sections, there are differing objectives for diameter growth between forest owners. Thus, adjustable growth models are needed to provide decision tools for controlling diameter growth for a wide range of possible objectives. In the following sections an allometric model describing the relation between crown width and diameter at breast height is presented. This model is then used to give quantitative information on production objectives.

Diameter growth at breast height is closely related to crown width development (e.g. (Curtis and Reukema 1970, Dawkins 1963, Hahn 1995, Hochbichler and Krapfenbauer 1988, Hummel 1951, Jobling and Pearce 1977, Krajicek et al. 1961, Lavny 2000, Mayer 1958, Mitchell 1969, Pryor 1988, Savill 1991, Spiecker 1983, Szappanos 1984). For further literature reviews see Hein (2004) and Hemery et al. (2005). The relation between these two parameters and tree age can be used as a basis for controlling diameter growth and for defining production aims. From the point of view of theoretical modeling, crown width is used as the dependent variable to predict the diameter at breast height, given that from the perspective of tree physiology assimilation is the driving factor for diameter growth of the trunk.

Data for this study were collected for common ash, sycamore and wild cherry in 13 countries (Hein 2004). The data cover a geographical area from 45° 30‘ to 55° 31‘ northern latitude and from - 2° 57‘ western to 18° 02‘ eastern longitude, which corresponds to an extension of 1500 km from west to east and 1300 km from north to south. Only dominant, co-dominant and sub-dominant trees were integrated into the data collection. No suppressed trees were measured. The methodology followed

Table 4.9. Range in data for the variables used in the model on crown width and for additional variables describing the data set.

variable	lower limit			upper limit			arith. mean			standard dev.		
	ash	sycamore	cherry	ash	sycamore	cherry	ash	sycamore	cherry	ash	sycamore	cherry
crown width: [m]	0.43	0.70	2.57	19.18	14.95	13.15	5.31	5.09	5.38	2.97	2.55	2.53
DBH: [cm]	3.5	3.4	10.6	96.5	88.8	62.2	28.2	28.2	25.5	16.36	15.37	15.39
tree age: [y]	11	10	18	205	179	83	50	54	43	32.05	32.51	22.91
tree height: [m]	7.4	4.6	9.7	45.8	37.8	34.6	23.1	21.2	15.72	7.71	6.46	5.79
mean radial increment: [mm]	0.78	1.07	1.76	6.5	8.46	6.06	3.0	2.88	3.92	0.94	0.98	0.74

Table 4.10. Results of the stepwise regression and the analysis of variance for common ash and sycamore for the variables of the crown width model, all parameter estimates significant at $\alpha = 0.0001$, wild cherry - age: significant at $\alpha = 0.005$; common ash: $r^2_{adj.} = 0.88$; sycamore: $r^2_{adj.} = 0.88$; wild cherry: $r^2_{adj.} = 0.93$.

variable	parameter estimate			standard error of the estimate			partial r^2			variance inflation factor		
	ash	sycamore	cherry	ash	sycamore	cherry	ash	sycamore	cherry	ash	sycamore	cherry
intercept	0.698	0.785	1.345	0.054	0.057	0.114	-	-	-	-	-	-
DBH: [cm]	0.203	0.183	0.183	0.003	0.003	0.009	0.86	0.87	0.92	3.3	3.5	5.9
tree age: [y]	-0.022	-0.016	-0.019	0.002	0.002	0.006	0.02	0.01	0.01*	3.3	3.5	5.9

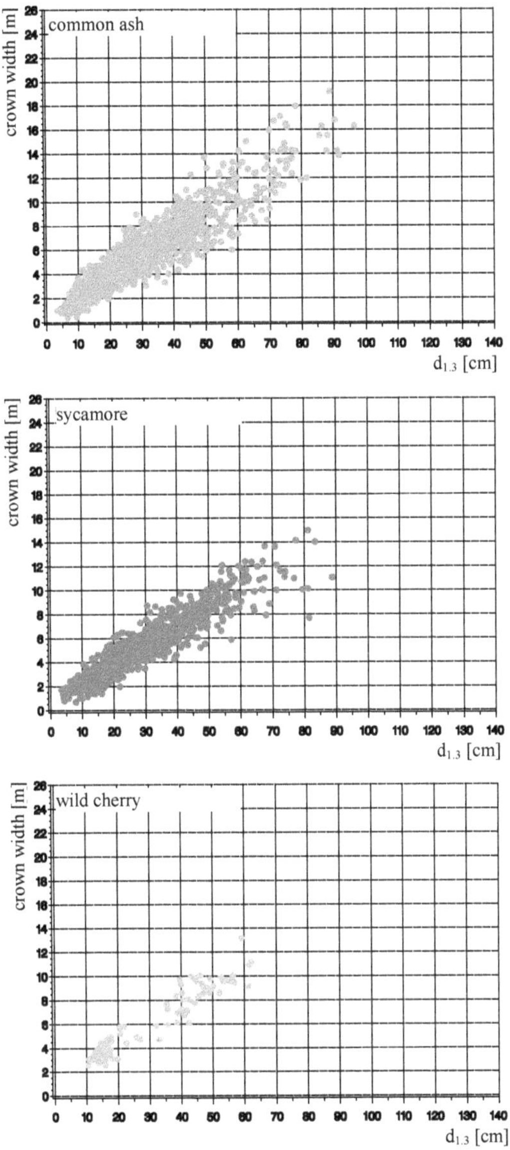

Figure 4.24. A strong correlation between crown width and diameter at breast height for common ash, as well as for sycamore and wild cherry can be recognized.

a retrospective approach, using the principle of space-for-time substitution with temporary plots of different developmental phases, with 25 trees in each plot. 1501 common ash, 1021 sycamore and 135 wild cherry trees with a wide range of tree dimensions were selected for the analysis (Table 4.9). As the sample for wild cherry trees is significantly smaller than those for the two other species and do not cover the lower range of dimensions and growth rates, results for wild cherry have to be considered as exploratory and preliminary and are subject to further validation.

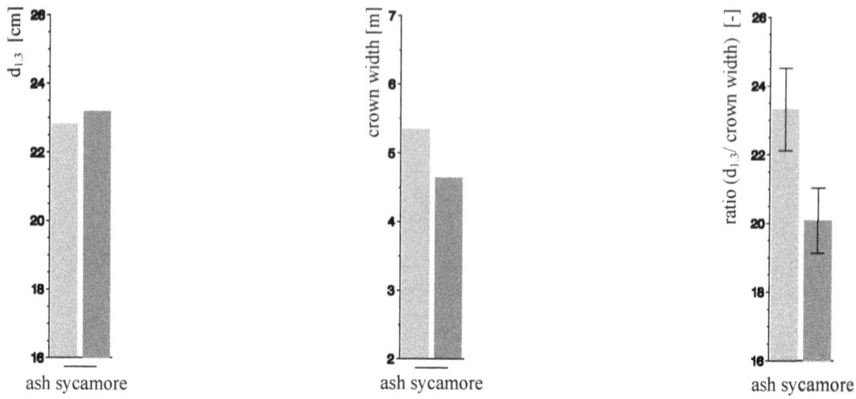

Figure 4.25. Comparison of the diameter at breast height (left) and crown width (middle) between ash and sycamore at age 33 showed no significant differences, Wilcoxon: $\alpha = 0.05$ while the ratio $d_{1.3}$/ crown width showed a significant difference (right, significant, T-test: $\alpha = 0.05$) of two pure 33 year old ash and sycamore stands. Vertical bar: confidence interval: 95%, N = 25 per stand. Not significant ($\alpha = 0.05$) differences are indicated by a linking horizontal bar.

A linear, multiple regression model was established to predict the crown width with diameter at breast height and tree age as independent variables (Table 4.10). With a stepwise regression, 88% of the total variation of crown width of common ash trees and sycamore could be explained (wild cherry: 93%). Most of the variation could be explained by the diameter at breast height. Tree age contributed only 1 to 2% to the reduction of the total variation. No significant interaction between the independent variables could be found. Also, the testing of co-linearity between the independent variables gave no indication of a severe mutual interdependence (VIF < 10). Indeed the VIF-value for the model for wild cherry may indicate a slightly higher interdependence due to an imbalanced data set (Belsley 1991, Draper and Smith 1998).

The visual check of the residuals did not reveal any sign of bias for trees from different classes of dominance and for trees with different mean radial increments. In addition, the model was tested to determine whether site productivity has a significant influence. As an indicator for site productivity, a site index using height at a base age previously developed by Hein (2004) was used. A significant influence of site quality on the equation was found for common ash and sycamore, but the contribution to the complete regression model was less than 1% and the impact on crown width was less than the presumed accuracy of the crown projection area measurement of 0.3 m. Therefore, site quality proved to have no influence relevant to forest practice on the relation between crown width, diameter at breast height and tree age. A similar result was found by Spiecker (1991) for oak.

The models are limited to a mean radial increment of 2 to 5 mm, a tree age ranging from 5 to 160 years and a diameter at breast height from 5 to 100 cm. In the case of wild cherry the model is limited to a diameter at breast height of 60 cm with a mean radial increment of 2 to 5 mm. Because there are only few observations for old trees with a high radial increment, a special model restriction is set: a radial increment of more than 5 mm at an age of 60 years and more than 4 mm at an age of more than 75 years for common ash and Sycamore.

To test whether there may be significant differences in crown width development between tree species, a corresponding test was carried out between common ash, sycamore and wild cherry. After pooling the data for common ash and sycamore respectively, and calculating a general linear model, a significant difference in the intercept between sycamore vs. cherry and cherry vs. ash was found. Another significant difference for the diameter at breast height – slope and for the age – slope between sycamore vs. ash was found. This indicates that for a given diameter at breast height, sycamore has a smaller crown width than common ash. A diameter at breast height of 60 cm at 75 years, which amounts to a mean radial increment of 4 mm, can result in a tree species dependent difference of 0.7 m crown width. Thus, sycamore with the same crown width as common ash can reach a slightly, but significantly, larger diameter at breast height. Unless further analysis on a larger dataset for wild cherry indicates otherwise, wild cherry trees seem to take a place between common ash and sycamore at an older age, but show larger crowns at young ages. To further test tree species dependent differences in crown width development, an additional comparison of diameter at breast height, crown width and the ratio ($d_{1.3}$/ crown width) was calculated.

Two separate pure stands of 25 common ash and sycamore trees neighboring each other were selected, measured and compared. The two stands were both established in 1968. Thinning was done on both plots in an identical manner. The mean diameter at breast height of common ash was slightly smaller than that of sycamore (Figure 4.25). The mean crown width was slightly larger than that of the neighbouring sycamore. Nevertheless these differences were not significant. There were, however, significant differences in the mean ratio ($d_{1.3}$/ crown width) between the pure ash and sycamore stand. Thus, there is a clear indication that for a given diameter at breast height sycamore shows a significantly smaller crown width than common ash. Using tree age as another independent variable in the crown width model, differences in crown width between fast and slow growing trees could be explained: fast growing trees (i.e. trees with a high mean radial increment) reach a certain diameter at breast height earlier than those with a relatively low mean radial increment. This diameter is reached with a significantly larger crown. This trend was confirmed for all three species. The same observations were described for oak (Spiecker 1991). The relation between radial increment, age and crown width is described in Figure 4.26. Common ash that grew at a mean radial increment of 5 mm per year reaches a diameter at breast height of 60 cm and a crown width of 11.6 m whereas trees growing at a rate of 2 mm reach the same diameter with a 0.9 m smaller crown width. The implications of differences in crown width for the production objectives are illustrated in the following section.

Production objectives

Different management objectives concerning diameter at breast height and mean radial increment result in differences in total production time and in the number of crop trees per hectare at the end of production. The length of the branch free bole is

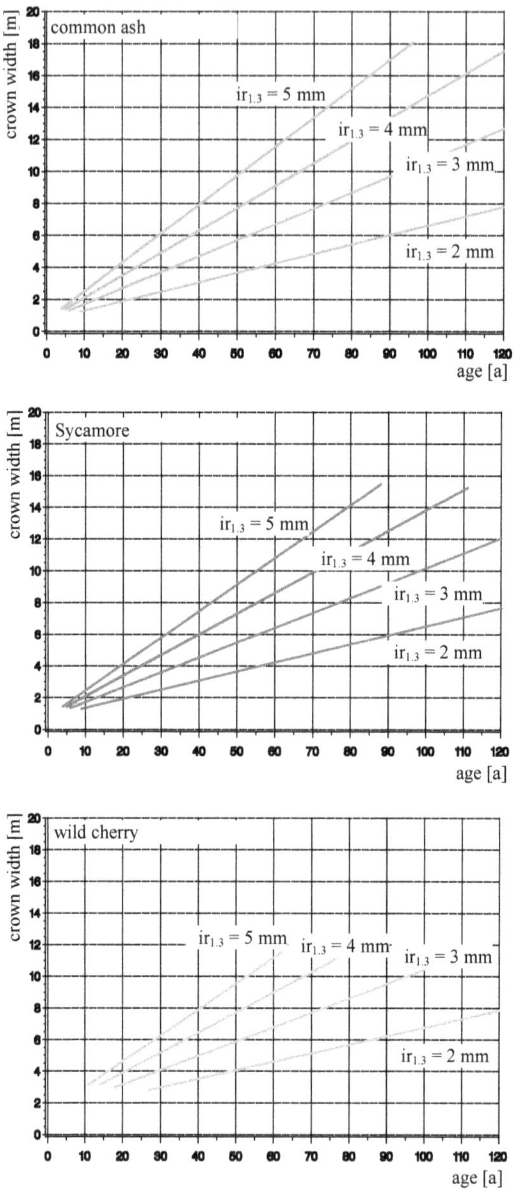

Figure 4.26. Modeled crown width for different levels of mean radial increment: 2, 3, 4, and 5 mm/ y for common ash, sycamore and wild cherry.

also affected. Table 4.11 presents a variety of production aims for common ash, sycamore and wild cherry and points out differences within and between the species. The calculation of production objectives requires assumptions on crown cover. The percentage of the crown cover was not investigated in this study and no

Table 4.11. Production objectives (harvesting diameter, production time, number of crop trees/ ha and the length of the clear bole) for common ash, sycamore, and wild cherry. Site index: common ash: 33 m at age 60 y; sycamore: 30 m at age 60 y; crown cover: 70% for common ash and sycamore and 50% for wild cherry. Less crown cover of wild cherry is needed to avoid die back of lower branches as these branches are very light demanding.

	harvesting diameter [cm]	mean radial increment [mm/ y]	Production time [y]	number crop trees/ ha [-]	length of the clear bole [m]
common ash	60	2	150	88	-
		3	100	71	20.3
		4	75	64	16.0
		5	60	61	12.8
	50	2	125	124	-
		3	83	100	19.3
		4	63	90	15.2
		5	50	85	12.0
sycamore	60	2	150	92	-
		3	100	78	14.4
		4	75	72	11.8
		5	60	69	9.7
	50	2	125	128	-
		3	83	109	14.2
		4	63	101	11.6
		5	50	97	9.6
wild cherry	60	2	150	-	-
		3	100	53	-
		4	75	49	-
		5	60	46	-
	50	2	125	-	-
		3	83	73	-
		4	63	67	-
		5	50	63	-

other results related to the species involved are available. It is assumed that the crown coverage is similar to the cover of oak which amounts according to investigations by Spiecker (1991) in average about 70%. A cover of 70% is applied to sycamore and common ash. Based on observations of wild cherry (Spiecker 1994) here a crown coverage of 50% is assumed. The indicated figures are within the limits of the model. For a complete comparison of differences in the natural pruning dynamics, the length of the clear bole is also presented. For instance, a harvesting diameter of 60 cm can be reached within 75 years, assuming a mean radial increment of 4 mm per year. Because of the specific crown width development of common ash, 64 mature crop trees per hectare are expected at the end of production time. If it is decided to grow trees at the lower mean radial increment of 3 mm, the production time is extended by 25 years and 71 common ash trees can grow on a hectare. With the same setting concerning harvesting diameter and a mean radial increment of 4 mm, 72 sycamore trees, or 78 trees for a scenario of 3 mm, can grow on an area of one hectare. In the case of wild cherry, 51 trees growing at 4 mm per year fit on one hectare or 55 trees with a mean radial increment of 3 mm. As cherry trees are light demanding and crowns of fast growing individuals are wider and longer than for slow growing individuals, the area not covered by their crowns has to be larger in order to avoid die back of lower branches. This means that the number of cherry crop trees indicated in the table need to be reduced for the fast growth scenarios (Spiecker 1994). The last column indicates the expected length of the clear bole for the site indices indicated above.

To obtain shares of crop trees in mixed stands, the figures for pure stands can be taken and multiplied with the desired share of the mixed species at the end of production time. When mixing wild cherry with tree species growing for a longer production time, it is possible to harvest wild cherry earlier leaving then space for the remaining trees.

What can we learn from open grown trees?

Open grown trees that grew without any competition from neighboring trees reveal the growth potential of a species under specific site conditions. With respect to diameter growth they show the maximum growth a tree can obtain under given environmental circumstances. As diameter growth is related to crown width, crown development also reaches its maximum. For this analysis, data of 42 open grown common ash and 32 open grown sycamore trees were collected in several European countries. Open grown trees were measured only on sites with productivity similar to that of forest sites. To ensure that the trees chosen had existed in a condition of predominantly free growth throughout their lives, trees with a first green branch higher than 5 m on any side were excluded. For this reason one-sided suppressed trees were also excluded. No open grown wild cherry were available. A more detailed description of the data is given in Table 4.12. The crown width model for trees from stands did not correctly predict the crown width of open grown trees. Crown width as indicated by the model presented in Table 4.10 systematically underestimated the crown width of solitary trees. For a given diameter at breast height, open grown trees exhibit a significantly larger crown width than trees from forest stands (cf. Figure 4.27). After a log-transformation of the dependent and the independent variables, a linear regression fitted the data best (Table 4.13).

A virtual stand of open grown trees can be calculated with the crown width model for open grown trees (cf. Table 4.14). This helps to determine the absolute minimum of production time and the number of crop trees per hectare at final stocking. Below the minimum number of crop trees at final stocking, losses of volume increment of valuable wood per hectare are expected. At a diameter at breast height of 60 cm, a crown width of 14.0 m and crown projection area of 154.1 m^2 can be expected for common ash. Assuming a crown cover of 70%, only 45 trees can grow in a virtual stand of one hectare size. Crown width of sycamore open grown trees is slightly smaller: a 60 cm tree covers a crown projection area of 125.7 m^2, which corresponds to a crown width of 12.7 m. Therefore 56 open grown sycamore trees are needed to cover a one hectare virtual stand.

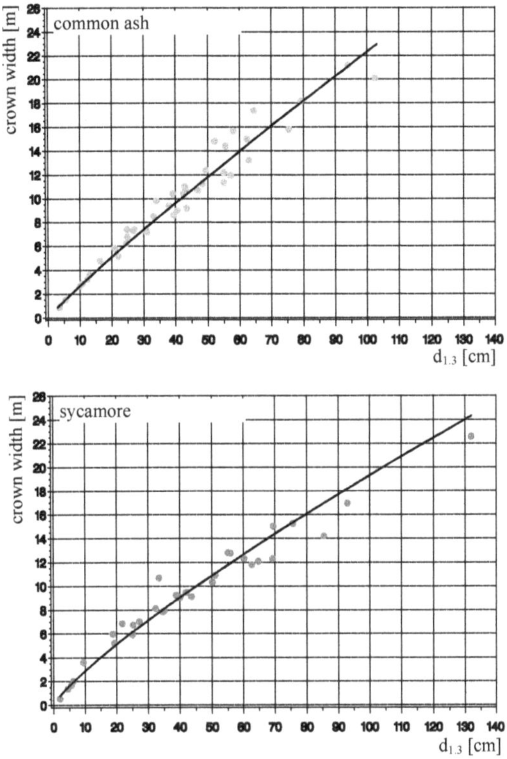

Figure 4.27. Crown width development of common ash and sycamore open grown trees; parameters of the regression line: cf. Table 4.13.

Assuming a mean radial increment of 7 mm for both species, a diameter at breast height of 60 cm will be reached in 43 years. According to previous results on pruning dynamics of open grown common ash and sycamore (cf. Hein 2004), a crown base height of 2.6 m in the former and 1.9 m in the latter mentioned species can be expected. This corresponds very well to results of Hasenauer (1997) in a previous study. An open grown ash tree with a height of 28 m will have a relative crown length of 90.7%, and sycamore trees a somewhat longer relative crown length at 92.7%.

4.3.3 Decision tools for controlling diameter growth

General principle of thinnings

Thinnings may increase the value of the logs without high investment. Efficient thinnings have to concentrate on the value production. They concentrate on the dominant and co-dominant tree layer. As most of the value is produced by the so

Table 4.12. Range of data for the variables used in the model on crown width for open grown trees and additional variables describing the data set.

variable	lower limit ash	lower limit sycamore	upper limit ash	upper limit sycamore	arithm. mean ash	arithm. mean sycamore	standard dev. ash	standard dev. sycamore
crown width: [m]	0.87	0.58	20.31	22.58	9.69	9.28	4.780	4.870
DBH: [cm]	3.5	2.0	102.5	132.0	40.9	43.4	22.74	29.04
tree age: [y]	7	5	169	162	49	58	37.06	38.20
tree height: [m]	3.4	2.3	30.4	23.0	13.5	11.7	5.15	5.20
mean radial increment: [mm]	1.48	1.85	8.27	6.5	4.67	3.62	1.62	1.36

Table 4.13. Results of the regression and the analysis of variance for common ash and sycamore for the variables of the crown width model for open grown trees, all parameters significant at $\alpha = 0.0001$; ln (crown width) = f(ln(DBH)).

parameter	estimate		standard error of the estimate		$r^2_{adj.}$	
variable	ash	sycamore	ash	sycamore	ash	sycamore
intercept	-1.119	-0.844	0.080	0.086	-	-
DBH: [cm]	0.918	0.826	0.022	0.024	0.98	0.98

called future crop trees, management activities have to concentrate on these trees. The value of the crop trees is determined by the volume and dimension of the clear bole. Therefore natural or artificial pruning and controlling crown expansion are key elements for controlling value production. Natural pruning and crown expansion are controlled by spacing and the intensity of thinning. For producing valuable clear boles of large dimension a two phase management system is recommended (Spiecker 1991). During the first phase pruning of the future crop trees is emphasized by natural pruning caused by competition or by artificial pruning. During the second phase diameter increment of the future crop trees is stimulated by repeated crown release. In this phase tree competition must be avoided so that lower branches do not die and crown base will stay at a constant height.

As height increment of valuable broadleaves generally culminates at rather early tree ages also natural pruning starts early and it then slows down when height increment slows down. Reaction of crown expansion after release is high when trees are young and height growth is fast and it generally decreases with age. Therefore controlling pruning and thinning activities have to start early. The thinning cycle may be shorter in the early phase of thinning and more trees per future crop tree need to be cut. The time at which to initiate phase two depends on the sensitivity of the log price to diameter of the log. When sensitivity is high phase two may start earlier.

Table 4.14. Quantitative description of common ash and sycamore trees in a virtual stand of open grown trees (crown cover: 70%); no wild cherry open grown trees were available.

	common ash	sycamore
Diameter at breast height: [cm]	60	60
Crown width: [m]	14.0	12.7
Crown projection area: [m]	154.1	125.7
Crop trees / ha:[-]	45	56
Mean radial increment: [mm]	7.0	7.0
Tree age: [y]	43	43
Height of first green branch: [m]	2.6	1.9
Tree height: [m]	28	26
Rel. crown length: [%]	90.7	92.7

Thinning is employed to reduce inter tree competition in order to release the crowns of the future crop trees. Increased space provides more resources for the tree such as sunlight, carbonic gases, nutrients and water. Selection of future crop trees may also help to achieve the desired species composition and stand structure. Criteria for selection of future crop trees include diversity and aesthetics in addition to quality and vigor. Permanent marking of future crop trees facilitates future thinning activities. Thinning is a tool for forming habitats for animals and plants. Litter decomposition may be stimulated by thinning which will allow more light to reach the ground. In the long run the height/diameter ratio may be lowered and the mechanical stability of the tree and the stand against storms, heavy snow and ice will increase. When crowns expand faster the diameter increment of the trunk is accelerated. By reducing the number of trees, growth is concentrated on the remaining trees especially on the high quality future crop trees. Their dimension and value increases faster and the production period may be shortened. Concentration on the release of the future crop trees may help to reduce costs by leaving the intermediate areas untouched. On the other hand a lack of thinning may result in a reduction of flora and fauna diversity and harvesting of low valued and small sized timber.

Concepts for controlling diameter growth

The decision tools presented here follow the concept of future crop tree selection. Decision criteria are the final diameter of the crop and production time, and thus also the mean radial increment and the length of the clear bole. The concept of the two phase thinning system is employed for growth control. During the first phase from stand establishment to the time when the desired length of the clear bole is reached, forest operations focus on natural or artificial pruning in order to reduce the width of the knotty core inside the trunk. The second phase concentrates all forest operations on speeding up diameter growth until the final harvesting diameter is reached. With this scenario a significant reduction of the knotty core can be obtained through a lower radial increment in earlier times of tree development and an almost unrestricted growth after a certain clear bole is reached. In contrast to other thinning types, the concept of crop tree selection thinning practice reduces the activities to the immediate surroundings of the selected future crop trees. Space between the crop trees is occupied by competitors or indifferent trees. The competitors are removed whenever they have a negative impact on future crop trees. Thus, a selective thinning is applied.

Future crop trees are selected when the desired clear bole length is reached. According to the figures given in chapter 4.2 on the best sites for common ash with a mean radial increment of 3 mm (for example a clear bole length of 10 m will be reached at a tree age of 27 years). Selection of crop trees should be done according to vitality, quality and distribution. Only trees that are vigorous should be selected, among these only those with the highest quality, and lastly the distribution of final crop trees should be taken into consideration. The number of future crop trees to be selected should not be larger than the number of mature final crop trees fitting present on the area. In case a future crop tree does not fulfill the expectations a neighboring reserve tree may take its position. On the other hand pruning and releasing of reserve trees causes additional cost. In addition reserve trees have to be removed before the final diameter is reached in order to prevent the diameter increment of the future crop trees from declining and running the risk of the crop trees not reaching their final diameter in time.

Because of differences in thinning practice in Europe, the following three sections show different decision tools for thinning control: firstly, a decision tool using the number of competitors per crop tree to be removed; secondly, there is a simple thinning rule using the number of remaining trees per hectare; and thirdly, there is a minimum-distance thinning rule. As the underlying relation of crown width, diameter at breast height and tree age is site independent, as described in the previous sections, all decision tools presented here are independent of site quality.

Number of competitors to be removed per future crop tree

The following thinning tool relies on a thinning in favour of individual trees and not on a hectare basis. The tool calculates the number of competitors per crop tree that have to be removed in order to maintain a desired level of radial increment.

The theoretical background for calculating the number of competitors to be removed was first described by Spiecker (1983). The main assumption is that all trees are distributed in a regular equilateral triangle. If there is only one crop tree around the competitor, the removal of the latter will give 1/6 of the space it occupied to the crop tree. With only crop trees remaining, 6/6 of the space of a removed competitor tree will be available to the benefit of the crop trees. Therefore, a different number of competitors have to be removed depending on the percentage of area occupied by crop trees. In these tables it is assumed that the space for crown expansion available to the crop trees will allow a constant diameter growth of the trunk.

For a target diameter of 60 cm and 65 common ash trees per hectare (sycamore: 75 trees), Table 4.15 and Table 4.16 show the number of competitors per crop tree to be removed within the next five years in order to maintain the tree's previous level of radial increment. For instance, for a 35 year old common ash crop tree with a diameter at breast height of 25 cm, 0.9 competitors have to be felled. Thus, for 10 crop trees with that diameter, 9 competitors have to be removed. With a diameter at breast height ratio between the crop tree and the competitor of less than 100%, more competitors have to be taken out because the crown width of the competitor is significantly smaller. To raise the radial increment of the crop tree to a higher level, more competitors must be removed. Because of the small dataset for wild cherry a corresponding table for this species has not yet been developed. The number of competitors per crop tree to be removed is a flexible tool to control growth as it takes into account the individual size and competition of the tree as well as different production objectives. However, this tool does not explicitly indicate the need for thinning. An indicator could be the diameter ratio between the crop tree and the competitor: the larger the diameter of the competitor in comparison to the diameter of the crop tree, the more urgent is the need for thinning.

In the case that the number of crop trees per hectare differs significantly from the number indicated in Tables 4.15 and 4.16, deviations from the number of competitors may be needed. From the moment when more than one crop tree benefits from the free space, the number of competitors per crop tree to be removed changes almost inversely proportional to the number of crop trees. Therefore, Table 4.17 presents the time from which more than one crop tree benefits from thinning.

Number of trees per hectare

In contrast to the decision tools described in the previous section, the number of trees per hectare is a tool that reflects the stand density on a hectare basis and does not allow for a decision according to the growing situation of the individual tree. Figure 4.28 displays the development of the stem number per hectare through time. Only dominant, co-dominant and sub-dominant trees and no suppressed trees are represented in the model.

As the development of the diameter at breast height is dependent on the crown width, in stands with trees growing at a low mean radial increment crown width is smaller and more trees will grow per hectare at a given age than in stands with trees growing faster in diameter. The number of trees per hectare can be easily calculated using the crown width model and assuming a certain crown cover (here: 70% for ash and sycamore or 50% for cherry).

For thinning practice the development of the number of trees as a function of age can be taken as a guideline, and thinning can be used to converge to the desired curve. In order to avoid losses of volume increment on a hectare base or losses in vitality or epicormics, it is recommended not to drastically reduce the number of trees at once.

Mean distance of neighbouring trees to the crop tree

Results of the previous decision tool can easily be transformed into a table of mean distances between trees for distinct scenarios of radial increment (Table 4.18). To

Table 4.15. Number of competitors per crop tree to be removed within the next 5 years for 65 final crop trees/ ha, a target diameter of 60 cm, and a diameter at breast height ratio between the crop tree and the competitor of 100%, 90% and 80%, crown cover 70% (common ash).

$d_{1.3}$ of the competitor = 100% of the $d_{1.3}$ of the crop tree																		
$d_{1.3}$ of the crop tree [cm]	age [years]																	
	15	20	25	30	35	40	45	50	55	60	65	70	75	80	85	90	95	100
10	3.2	2.3	1.7															
15	3.6	2.6	2.0	1.6	1.4													
20		2.1	1.7	1.5	1.3	1.2	1.1	1.0										
25			1.1	1.0	0.9	0.8	0.7	0.7	0.6	0.6								
30				0.7	0.6	0.6	0.5	0.5	0.4	0.4	0.4	0.4	0.4					
35					0.5	0.4	0.4	0.3	0.3	0.3	0.3	0.3	0.3	0.3	0.3			
40						0.3	0.3	0.3	0.2	0.2	0.2	0.2	0.2	0.2	0.2	0.2	0.2	0.2
45							0.2	0.2	0.2	0.2	0.2	0.2	0.2	0.1	0.1	0.1	0.1	0.1
50								0.2	0.2	0.2	0.1	0.1	0.1	0.1	0.1	0.1	0.1	0.1
55									0.2	0.2	0.1	0.1	0.1	0.1	0.1	0.1	0.1	0.1
60										0.2	0.1	0.1	0.1	0.1	0.1	0.1	0.1	0.1

$d_{1.3}$ of the competitor = 90% of the $d_{1.3}$ of the crop tree																		
$d_{1.3}$ of the crop tree [cm]	age [years]																	
	15	20	25	30	35	40	45	50	55	60	65	70	75	80	85	90	95	100
10	3.8	2.7	2.1															
15	4.3	3.2	2.5	2.0	1.7													
20		2.5	2.1	1.8	1.6	1.5	1.4	1.3										
25			1.4	1.2	1.1	1.0	0.9	0.8	0.8	0.8								
30				0.9	0.8	0.7	0.6	0.6	0.6	0.5	0.5	0.5	0.5					
35					0.6	0.5	0.5	0.4	0.4	0.4	0.4	0.4	0.3	0.3	0.3			
40						0.4	0.4	0.3	0.3	0.3	0.3	0.3	0.3	0.2	0.2	0.2	0.2	0.2
45							0.3	0.3	0.2	0.2	0.2	0.2	0.2	0.2	0.2	0.2	0.2	0.2
50								0.2	0.2	0.2	0.2	0.2	0.2	0.2	0.1	0.1	0.1	0.1
55									0.2	0.2	0.2	0.2	0.2	0.2	0.1	0.1	0.1	0.1
60										0.2	0.2	0.2	0.2	0.2	0.1	0.1	0.1	0.1

$d_{1.3}$ of the competitor = 80% of the $d_{1.3}$ of the crop tree																		
$d_{1.3}$ of the crop tree [cm]	age [years]																	
	15	20	25	30	35	40	45	50	55	60	65	70	75	80	85	90	95	100
10	4.6	3.3	2.6															
15	5.3	3.9	3.1	2.6	2.2													
20		3.1	2.6	2.3	2.1	1.9	1.8	1.7										
25			1.8	1.5	1.4	1.2	1.2	1.1	1.1	1.0								
30				1.1	1.0	0.9	0.8	0.8	0.7	0.7	0.7	0.7	0.6					
35					0.7	0.7	0.6	0.6	0.5	0.5	0.5	0.5	0.5	0.4	0.4			
40						0.5	0.5	0.4	0.4	0.4	0.4	0.3	0.3	0.3	0.3	0.3	0.3	0.3
45							0.4	0.3	0.3	0.3	0.3	0.3	0.3	0.2	0.2	0.2	0.2	0.2
50								0.3	0.3	0.3	0.2	0.2	0.2	0.2	0.2	0.2	0.2	0.2
55									0.3	0.3	0.2	0.2	0.2	0.2	0.2	0.2	0.2	0.2
60										0.3	0.2	0.2	0.2	0.2	0.2	0.2	0.2	0.2

Table 4.16. Number of competitors per crop tree to be removed within the next 5 years for 75 final crop trees/ ha, a target diameter of 60 cm, and a diameter at breast height ratio between the crop tree and the competitor of 100%, 90% and 80%, crown cover 70% (sycamore).

$d_{1.3}$ of the competitor = 100% of the $d_{1.3}$ of the crop tree																		
$d_{1.3}$ of the crop tree [cm]	age [years]																	
	15	20	25	30	35	40	45	50	55	60	65	70	75	80	85	90	95	100
10	3.0	2.1	1.6															
15	3.4	2.5	1.9	1.6	1.3													
20		1.9	1.5	1.3	1.1	1.0	0.9	0.9										
25			1.1	0.9	0.8	0.7	0.6	0.6	0.6	0.5								
30				0.7	0.6	0.5	0.5	0.4	0.4	0.4	0.4	0.3	0.3					
35					0.4	0.4	0.4	0.3	0.3	0.3	0.3	0.3	0.2	0.2	0.2			
40						0.3	0.3	0.3	0.2	0.2	0.2	0.2	0.2	0.2	0.2	0.2	0.2	0.2
45							0.2	0.2	0.2	0.2	0.2	0.2	0.1	0.1	0.1	0.1	0.1	0.1
50								0.2	0.2	0.2	0.1	0.1	0.1	0.1	0.1	0.1	0.1	0.1
55									0.2	0.2	0.1	0.1	0.1	0.1	0.1	0.1	0.1	0.1
60										0.2	0.1	0.1	0.1	0.1	0.1	0.1	0.1	0.1

$d_{1.3}$ of the competitor = 90% of the $d_{1.3}$ of the crop tree																		
$d_{1.3}$ of the crop tree [cm]	age [years]																	
	15	20	25	30	35	40	45	50	55	60	65	70	75	80	85	90	95	100
10	3.5	2.5	2.0															
15	4.1	3.0	2.3	1.9	1.6													
20		2.3	1.8	1.6	1.4	1.3	1.2	1.1										
25			1.3	1.1	1.0	0.9	0.8	0.7	0.7	0.7								
30				0.8	0.7	0.6	0.6	0.5	0.5	0.5	0.4	0.4	0.4					
35					0.5	0.5	0.4	0.4	0.4	0.3	0.3	0.3	0.3	0.3	0.3			
40						0.4	0.3	0.3	0.3	0.3	0.3	0.2	0.2	0.2	0.2	0.2	0.2	0.2
45							0.3	0.2	0.2	0.2	0.2	0.2	0.2	0.2	0.2	0.2	0.2	0.2
50								0.2	0.2	0.2	0.2	0.2	0.2	0.1	0.1	0.1	0.1	0.1
55									0.2	0.2	0.2	0.2	0.2	0.1	0.1	0.1	0.1	0.1
60										0.2	0.2	0.2	0.2	0.1	0.1	0.1	0.1	0.1

$d_{1.3}$ of the competitor = 80% of the $d_{1.3}$ of the crop tree																		
$d_{1.3}$ of the crop tree [cm]	age [years]																	
	15	20	25	30	35	40	45	50	55	60	65	70	75	80	85	90	95	100
10	4.2	3.0	2.4															
15	4.9	3.6	2.8	2.3	2.0													
20		2.8	2.3	2.0	1.7	1.6	1.5	1.4										
25			1.6	1.4	1.2	1.1	1.0	0.9	0.9	0.8								
30				1.0	0.9	0.8	0.7	0.7	0.6	0.6	0.6	0.5	0.5					
35					0.7	0.6	0.5	0.5	0.5	0.4	0.4	0.4	0.4	0.4	0.4			
40						0.5	0.4	0.4	0.4	0.3	0.3	0.3	0.3	0.3	0.3	0.3	0.3	0.3
45							0.3	0.3	0.3	0.3	0.3	0.2	0.2	0.2	0.2	0.2	0.2	0.2
50								0.3	0.3	0.2	0.2	0.2	0.2	0.2	0.2	0.2	0.2	0.2
55									0.3	0.3	0.2	0.2	0.2	0.2	0.2	0.2	0.2	0.2
60										0.3	0.2	0.2	0.2	0.2	0.2	0.2	0.2	0.2

Table 4.17. Time from which more than one crop tree benefits from thinning of competitors (crown cover: 70%).

Mean radial increment	common ash			sycamore		
	at 60 crop trees/ ha	at 80 crop trees/ ha	at 100 crop trees/ ha	at 60 crop trees/ ha	at 80 crop trees/ ha	at 100 crop trees/ ha
	tree age [y]			tree age [y]		
2	54	45	39	54	45	39
3	32	27	24	33	28	24
4	23	19	17	24	20	17
5	18	15	13	19	16	14

maintain a certain level of radial increment with increasing diameter at breast height, an increasing distance between the crop tree and the competitors is necessary. For instance, in the case of a common ash tree with a diameter at breast height of 30 cm and a mean radial increment of 3 mm, a mean distance of 6.8 m is needed to maintain this level of diameter growth. Speeding up diameter growth to a higher level would require a greater distance.

Whereas this thinning rule gives quick insight into the growing space needed, there is no indication about the duration of growth reaction after the thinning. It is recommended not to release the subject trees to heavy thinning at once, as this may result in losses in vitality or quality.

An even simpler thinning rule can be calculated using a constant derived from the crown width model and assuming a crown cover of 70% for common ash and sycamore and 50% for wild cherry (Table 4.19). The constant is calculated as the arithmetic mean for a tree of 30 and 60 cm diameter at breast height. In the case of common ash with a mean radial increment of 4 to 5 mm per year, the diameter at breast height of a crop tree of 30 cm can be multiplied by a constant with the value 23. The result gives the required distance (approximated) between the crop tree and the nearest competitor to maintain a radial increment of 4 to 5 mm. Within a circle of 6.9 m all competitors have to be removed to ensure the desired level of diameter growth. For wild cherry Spiecker and Spiecker (1988) recommended a constant with a value 25. A similar decision tool has been developed by Armand (1995). His results agree with the results presented here.

Crown base for controlling thinning intensity

After the pruning process has been advanced to the desired height of the crown base diameter growth is stimulated by repeated releases of future crop trees. As from this time on, the crown base should not move further upward; thinning should be intense enough to allow for sufficient light incidence on the lower branches. This very simple and easy to apply rule has been recommended by Spiecker and Spiecker (1988) and has been applied successfully at various locations (see also Wilhelm and Raffel 1993).

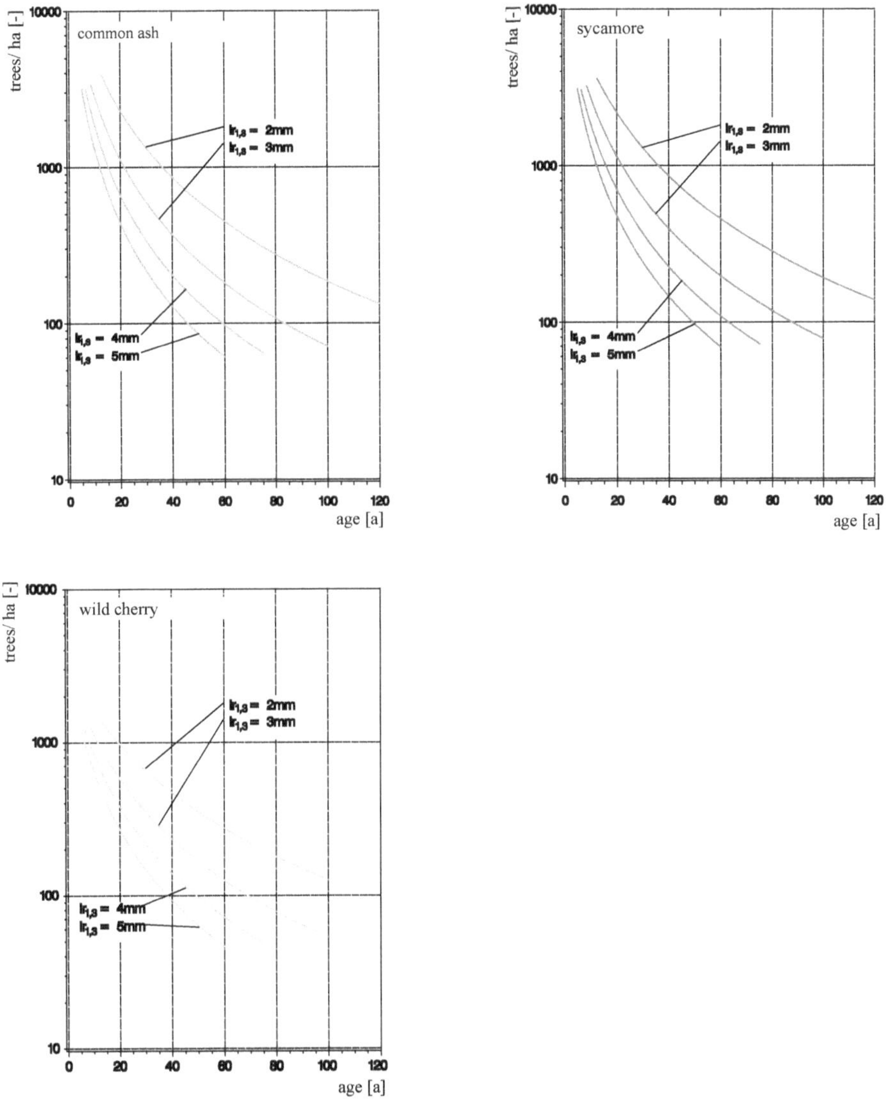

Figure 4.28. The reduction in number of trees per hectare as a function tree age is described for different scenarios of constant radial increment of common ash, sycamore and wild cherry trees (crown cover: 70% for ash and sycamore and 50% for cherry).

4.3.4 Conclusions

There is no one single ideal rule for the control of diameter growth. First, objectives concerning the target diameter depend very much on the growing conditions of the forest and preferences of the decision maker. Secondly, many decision rules could

Table 4.18. Mean distance between common ash and sycamore trees for different scenarios of radial increment and diameter at breast height of the crop trees (crown cover: 70% for ash and sycamore and 50% for cherry).

	common ash Mean distance between trees [m]										
Ø $ir_{1,3}$ [mm]	$d_{1,3}$ = 10 cm	$d_{1,3}$ = 15 cm	$d_{1,3}$ = 20 cm	$d_{1,3}$ = 25 cm	$d_{1,3}$ = 30 cm	$d_{1,3}$ = 35 cm	$d_{1,3}$ = m 40 c	$d_{1,3}$ = 45 cm	$d_{1,3}$ = 50 cm	$d_{1,3}$ = 55 cm	$d_{1,3}$ = 60 cm
2	2.6	3.5	4.4	5.2	6.1	7.0	7.9	8.8	9.7	10.5	11.4
3	2.8	3.8	4.8	5.8	6.8	7.8	8.8	9.7	10.8	11.7	12.8
4	2.9	4.0	5.0	6.1	7.1	8.2	9.2	10.3	11.2	12.4	13.3
5	3.0	4.1	5.2	6.2	7.3	8.4	9.5	10.6	11.7	12.7	13.8

	sycamore Mean distance between trees [m]										
Ø ir1,3 [mm]	$d_{1,3}$ = 10 cm	$d_{1,3}$ = 15 cm	$d_{1,3}$ = 20 cm	$d_{1,3}$ = 25 cm	$d_{1,3}$ = 30 cm	$d_{1,3}$ = 35 cm	$d_{1,3}$ = 40 cm	$d_{1,3}$ = 4 5 cm	$d_{1,3}$ = 50 cm	$d_{1,3}$ = 55 cm	$d_{1,3}$ = 60 cm
2	2.6	3.5	4.4	5.2	6.1	6.9	7.8	8.6	9.5	10.3	11.2
3	2.8	3.7	4.7	5.6	6.5	7.5	8.4	9.3	10.3	11.2	12.1
4	2.9	3.9	4.8	5.8	6.8	7.8	8.7	9.7	10.7	11.7	12.6
5	2.9	3.9	4.9	5.9	6.9	7.9	8.9	9.9	10.9	11.9	12.9

	wild cherry Mean distance between trees [m]										
Ø ir1,3 [mm]	$d_{1,3}$ = 10 cm	$d_{1,3}$ = 15 cm	$d_{1,3}$ = 20 cm	$d_{1,3}$ = 25 cm	$d_{1,3}$ = 30 cm	$d_{1,3}$ = 35 cm	$d_{1,3}$ = 40 cm	$d_{1,3}$ = 45 cm	$d_{1,3}$ = 50 cm	$d_{1,3}$ = 55 cm	$d_{1,3}$ = 60 cm
2	3.8	4.8	5.7	6.7	7.7	8.6	9.6	10.5	11.5	12.4	13.4
3	4.0	5.1	6.2	7.3	8.3	9.4	10.5	11.5	12.6	13.7	14.7
4	4.2	5.3	6.4	7.5	8.7	9.8	10.9	12.0	13.2	14.3	15.4
5	4.2	5.4	6.5	7.7	8.9	10.0	11.2	12.3	13.5	14.7	15.8

Table 4.19. Rule of thumb for the calculation of the mean distance between a crop tree and its neighbors for common ash, sycamore and wild cherry, crown cover: 70% for ash and sycamore and 50% for cherry.

Mean radial increment [cm]	$d_{1.3}$ [cm] × constant = distance [cm] common ash	sycamore	wild cherry
2–3	d1.3 × 21	d1.3 × 20	d1.3 × 25
4–5	d1.3 × 23	d1.3 × 22	d1.3 × 28

be established but probably none would satisfy all requirements of the forest owner, be appropriate to every type of stand structure or can be based on scientific analysis.

Even though the models presented here allow for the calculation of production objectives, especially of the targeted diameter in relation to the number of crop trees in the stand and the production time, and although they propose a variety of decision tools for thinning to reach these objectives, some questions still have to be answered.

Growth reaction to crown release after thinning may be different depending on the vitality of the trees or environmental conditions. More vital or dominant trees will probably react differently than formerly suppressed trees.

Even though trees from pure and mixed stands showed no significant difference in the crown width model, growth control of common ash, sycamore and wild cherry in mixture with other species should consider differences in height growth, shade tolerance and the potential to occupy growing space between the species.

The comparison made between common ash and sycamore concerning crown width indicates slightly smaller crowns for sycamore. A similar comparison with crown development of wild cherry was not possible because of insufficient data on this species. Such a comparison can only be conducted by considering growing time, i.e. tree age, and therefore the level of radial increment on a similar site. Further studies could include other species into such a comparison in order to identify species specific growth patterns in crown width and their expansion capacity.

In the case of common ash, sycamore and wild cherry, large radial increments do not compromise good mechanical properties (cf. Chapter 4.5 in this book). Thus, it seems to be appropriate to grow valuable timber in a short production time with large tree rings. Even though there were fast growing trees with a mean radial increment of more than 5 mm at a higher tree age in the data set, their share was smaller than that of young fast growing trees. Additional work needs to be done to explore the capacity of old trees that grew quickly in their youth to maintain that level of diameter growth even at older ages and larger diameters.

References

Armand, G. 1995. Feuillus précieux – Conduite des plantations en ambiance forestière – Mersisier, érabel sycomore, frêne, chêne rouge d'Amérique. (Valuable Broadleaves – Silviculture of Plantations on Forest Sites – Wild Cherry, Sycomore Maple, Ash and American Red Oak). Editions IDF, Institut pour le Developpement Forestier, Paris. (Original in French)

Barret, J.W. and Holmsgaard, E. 1964. Thinning in Hardwoods – Danish Guidelines for American Practices. Journal of Forestry 62: 716–720.

Belsley, D.A. 1991. Condition Diagnostics: Collinearity and Weak Data in Regression. Wiley Series in Probability and Mathematical Statistics, New York.

Curtis, R.O. and Reukema, D.L. 1970. Crown Development and Site Estimates in a Douglas-fir Plantation Spacing Test. Forest Science 16: 287–301.

Dawkins, H.C. 1963. Crown Diameters: Their Relation to Bole Diameter in Tropical Forest Trees. Commonwealth Forest Revue 42: 318–333.

Draper, N.R. and Smith, H. 1998. Applied Regression Analysis. Willey, 3rd Edition, New York.

Hahn, D. 1995. Entscheidungshilfe für die Steuerung des Dickenwachstums von Buche *Fagus sylvatica* L.). Entwicklung und Aufbau auf der Grundlage eines distanzabhängigen Einzelbaum-Konkurrenzmodells. (Decision Support for Controlling Diameter Growth of Beech (*Fagus sylvatica* L.) Based upon a Spatial Explicit Competition Model). Ph.D., Universität Freiburg i. Br.. (Original in German)

Hasenauer, H. 1997. Dimensional Relationships of Open-grown Trees in Austria. Forest Ecology and Management 96: 197–206.

Hein, S. 2004. Zur Steuerung von Astreinigung und Dickenwachstum bei Esche (*Fraxinus excelsior* L.) und Ahorn (*Acer pseudoplatanus* L.). Controlling Natural Pruning and Diameter Growth with Common Ash (*Fraxinus excelsior* L.) and Sycomore Maple (*Acer pseudoplatanus* L.). Ph.D. Universität Freiburg. (Original in German)

Hemery,G.E., Savill, P.S. and Pryor, S.N. 2005. Applications of the crown diameter-stem diameter relationship for different species of broadleaved trees. Forest Ecology and Management 215: 285–294.

Hochbichler, E. and Krapfenbauer, A. 1988. Behandlungsprogramme für die Werteichenproduktion im Wienerwald und Weinviertel. (Silvicultural Guide for the Production of Valuable Oaks in the Wienerwald and Weinviertel). Centralblatt für das gesamte Forstwesen 105: 1–23. (Original in German)

Hummel, F. C. 1951. Studies of Growth and Yield – increment of Free Grown Oak. Forestry Commission Report on Forest Research 12: 65–66.

Jobling, J. and Pearce, M.L. 1977. Free Growth of Oak. Forestry Commission Forest Record 113. Forestry Commission, London.

Kerr, G. and Evans, J. 1993. Growing Broadleaves for Timber. Forestry Commission Handbook 9.

Krajicek, J.E.; Brankmann, K.A. and Gingrich, S.F. 1961. Crown Competition - a Measure of Density. Forest Science 7: 35–42.

Lavny, V. 2000. On the Structure of Stands with Common Ash in the Western Forest Steppe of the Ukraine. Ph.D., Ukrainian State Foresttechnical University, Lviv. (Original in Ukraine)

Mayer, R. 1958. Crown Size and Increment of Sessile Oak on South-German Sites. Allgemeine Forst- und Jagdzeitung 129: 151–163. (Original in German)

Mitchell, K.J. 1969. Simulation of the Growth of Even-aged Stands of White Spruce. School of Forestry Bulletin 75. Yale University.

Pilard-Landeau, B. and Le Goff, N. 1996. Silviculture of Ash. Office National des Forêts (eds.), Bulletin technique 3: 9–14. (Original in French)

Pryor, S.N. 1988. The Silviculture and Yield of Wild Cherry. Forestry Commission Bulletin 75.

Savill, P.S. 1991. The Silviculture of Trees Used in British Forestry. Wallingford. 129 p.

Soulères, G. 1997. Valuable Broadleaves – Round Wood Prices and their Evolution 1955–1995. Forêts de France 404: 2–8. (Original in French)

Spiecker, H. 1983. Durchforstungsansätze bei Eiche unter besonderer Berücksichtigung des Dickenwachstums. (Thinning Guidelines with Oaks with Special Regard to Diameter Growth). Allgemeine Forst- und Jagdzeitung 154: 21–36. (Original in German)

Spiecker, H. 1991. Zur Steuerung des Dickenwachstums und der Astreiningung von Stiel- (*Quercus robur* L.) und Traubeneiche (*Quercus petrea* (Matt.) Liebl.). (On the Controlling of Diameter Growth and Natural Pruning of Sessile and Pedunculate Oak (*Quercus petrea* (Matt.) Liebl. and *Quercus robur* L.)). Schriftenreihe der Landesforstverwaltung Baden-Württemberg 72. Stuttgart. (Original in German)

Spiecker, H. 2001. Reforestation with Valuable Broadleaves after Storm. Freiburger Forstliche Forschung Berichte 25: 89–100. (Original in German)

Spiecker, H. 2003. Laubholzerziehung und Wertleistungsgrundsätze. (Management of valuable broadleaves and principles of value production). Österreichische Forstzeitung 114: 10–11. (Original in German)

Spiecker, M. and Spiecker, H. 1988. Erziehung von Kirschenwertholz (Management of wild cherry for valuable timber production). AFZ (20): 562–565 (Original in German)

Spiecker, M. 1994. Wachstum und Erziehung wertvoller Waldkirschen (Growth and Silviculture of Valuable Wild Cherry Trees). Mitteilungen der Forstlichen Versuchs- und Forschungsanstalt Baden-Württemberg 181. Original in German)

Szappanos, A. 1984. Richtlinien für die Züchtung von E-Bäumen der Traubeneiche. (Guidelines for Growing Crop Trees with Sessile Oak). Acta Facultatis Forestalis 1: 31–50. (Original in German)

Whiteman, A.; Inley, H. and Watt, G. 1991. Price-Size Curves for Broadleaves. Forestry Commission Occasional Paper 32. Edinburgh.

Wilhelm, G.J. and Raffel, D. 1993. La sylviculture du mélange temporaire hetre-merisier sur le Plateau Lorrain. (Silviculture of Time-Limited Mixtures of Wild Cherry and Beech on the Lorrain Plateau). Revue Forestière Française XLV(6): 651–668. (Original in French)

4.4 Final Cutting Systems of Valuable Broadleaves

Noël Lust
Department of Forest and Water Management, Laboratory of Forestry, Ghent University, Belgium

Abstract

This chapter mainly addresses three issues: 1. the size of the final cutting unit, 2. the duration of the final cutting and 3. the rotation period. Ash, sycamore and wild cherry usually occur on small areas, intimately or group-mixed. During the last decades, however, small monocultures were also planted. Three typical final cutting systems are distinguished: single tree cutting, group cutting and small stand cutting. It is evident that final cuttings often occur on a single tree level. A group cutting is applied in group-mixed stands. Duration of the final cuttings depends on several factors, but periods of 10 to 20 years are generally recommended.

Most authors suggest a rotation period of 60–80 years. The period is determined by site characteristics, growth potentials, production goals and timber quality. Information on other valuable broadleaves is extremely scarce.

Keywords: final cutting system, valuable broadleaves, ash, sycamore, wild cherry

4.4.1 Introduction

The keyword "final cutting system" is only rarely found in literature and certainly not with respect to the valuable broadleaves common ash (*Fraxinus excelsior* L.), sycamore (*Acer pseudoplatanus* L.) and wild cherry (*Prunus avium* L.). Authors such as Joyce (1998), Schütz (1997), Matthews (1989), Burschel and Huss (1987), Mayer (1977) and Dengler (1971) do not mention it.

The term "final cutting system" partially refers to and overlaps with the well known term "silvicultural system". Although the former term is not scientifically defined, it can be considered more restrictive than the latter.

From a silvicultural point of view, five different silvicultural systems are normally distinguished, i.e. clear-felling, strip felling, shelterwood felling, group felling and selective felling. They are characterized by the global regeneration method. Final cutting systems, on the contrary, are mainly characterised by the size and the duration of the fellings.

In this paper the following two issues will be discussed: 1. size of the final cutting unit and 2. duration of the final cutting. In addition to these major questions, and almost of the same importance, the topic of rotation will be discussed. Hardly any silvicultural data pertaining to final cutting systems of valuable broadleaves are available whereas much data can be found regarding rotation periods.

Final cutting can occur on both small and large scales. Cutting can be executed all at once or spread over several years, including decades. The size of felling varies from the individual tree, featured by crown width, to an indefinite large space. As a sort of compromise between a large clear-felled area and a single tree gap, group fellings are presently slowly dominating the European final cutting systems. It is likely that such systems will become even more important in the future because of the increasing demand for sustainable forestry management practices. According to authors and forestry schools, group fellings vary in size, although there is a large consensus to fix the group size to areas ranging between 0.1 and 0.50 ha. Regarding the valuable hardwood species, clear-felling of large areas does not occur although a fourth size class, between a clear-felled area and a group, may be considered as an option. We call it a small stand with an area of approximately one hectare.

In the discussion of final cutting the rotation period is undoubtedly an important issue. The term rotation is used to describe the interval between successive crop regenerations (Savill et al. 1997). In the past, rotations were normally fixed in the management plan and had to be strictly respected. However, rotations are presently generally determined by a variety of factors such as financial returns, maximal volume yields, a specified log size, silvicultural reasons, nature conservation concerns, unforeseen events such as diseases, storms and market changes, the objectives and position of the owner, site and stand improvements by fertilizers, tree breeding or general silvicultural treatments, etc. In any case, the idea of a fixed rotation should be abandoned as it is an obstacle to the correct choice of the most suitable felling time.

The lack of references related to the cutting system of the discussed species is not surprising as these species do not occupy large areas and typical final cutting systems are not conducive to their natural distribution. Generally speaking, these species appear on small areas, intimately or group-mixed with other hardwoods. They mainly occur in mixtures with beech (*Fagus sylvatica* L.) or oak (*Quercus spp.*) as the major tree species; two species which have totally different silvicultural characteristics. Because of this, final cuttings are not determined by the rarely appearing valuable broadleaves, but by the more common long-lived European hardwoods which are managed according to their own specific final cutting systems.

Of additional significance is the presence of valuable broadleaves in the traditional coppice with standards system. In these forests mainly comprised of oak, valuable broadleaves are uneven-aged and normally occur in an intimate mixture with other hardwoods. Last century most of the coppice with standards forests were transformed into 0.5–1.0 ha high forest systems characterised by an intimate or group mixture and a low density. This forest type sometimes developed into a kind of selection forest (futaie claire) (Boudru 1986, Lanier 1986) characterised by uneven-aged trees that are individually mixed. Such an occurrence implies that typical final cutting systems can not be applied and that fellings must occur on an individual level.

Ash, maple and cherry do, however, also naturally appear in associations where they dominate, and where beech or oak do not, and eventually mix with species such as elm (*Ulmus spp.)* and lime (*Tilia spp.*). Such stands, however, only occur on specific sites which feature very high quality in terms of both nutrients and moisture.

The common natural range of these valuable broadleaved stands accounts for a number of identical silvicultural characteristics. Consequently the final cutting systems are in many cases similar for these species. Nevertheless, a number of differences between these species do exist. Differences in growth characteristics also mean that the objectives, including production goals, are not always the same. Sycamore often has a principally cultural value (Neirynck et al. 2000), e.g. in the conversion of large scale homogeneous beech stands, or even Norway spruce (*Picea excelsa* L.) stands (Kazda and Pichler 1998).

Over the last decades, however, the high specific value of valuable broadleaves has become increasingly recognised. As a result, also the silvicultural use of these species has changed. Instead of in intimate mixtures, the species has increasingly been planted and managed in groups; on certain sites pure stands of ash, maple or cherry monocultures have been established. Valuable broadleaves have also become important species in the frame of biodiversity enhancement and afforestation of abandoned agricultural lands (Christensen and Emborg 1996). All these practices automatically lead to typical final cutting systems.

Furthermore, basic differences in silvicultural visions and management systems between different forest regions exist and correspond to some silvicultural traditions. Stand density, tending intensity, silvicultural goals and practices often display noticeable differences between central European countries, Boreal countries, the United Kingdom and France.

Although ash, sycamore and wild cherry have much in common in terms of site requirements, growth rate and silvicultural treatment (Franc et al. 1992, Houillier and Rittié 1992, Revue 1992), some specific issues are worthwhile to underline with respect of the final cutting systems.

Ash is more suitable in mixtures with other broadleaved species than in pure plantations (Evans 2001, Richter 2001, Joyce 1998). Site requirements of ash are so specific that the area on which a pure stand can successfully be grown is often limited (Lévy et al. 1992). In permanent mixtures with beech, oak and sycamore, it should be taken into consideration that the rotation for ash and sycamore is similar, whereas beech and oak have rotations that are much longer. In most broadleaved woodlands ash regenerates in small openings where it forms dense thickets.

Like common ash, sycamore also forms a component of broadleaved woodlands and is rarely found in pure stands. However, sycamore occupies a wider range of site conditions than ash and therefore pure stands are more likely to occur. In mixtures with its natural companion beech, sycamore will only compete up to approximately 50 years of age. Therefore intimate mixtures of sycamore and beech should not be favoured.

Wild cherry normally occurs as an occasional tree species or in small groups in mixed woodlands. It has a relatively short lifespan and propagates rapidly by root suckers. Mixtures with oak are less problematic than with beech, which is a competitor from mid-rotation and thereafter. The less competitive the cherry, relative to the beech on a particular site, the larger the groups of cherry can be (Joyce 1998, Franc and Ruchaud 1996, Franc et al. 1992). Over the last decades, mainly due to the very high prices of wild cherry timber, many owners planted the species in small pure stands of 0.5–1.0 ha, e.g. on former agricultural land along with ash and maple (Schrötter 2001).

4.4.2 Final cutting systems

Valuable broadleaves are managed in various ways in accordance with their natural or artificial occurrence in the stand; single tree cutting, group cutting and small stand cutting are the most common cutting systems. Traditional strip cutting, shelterwood and typical selection cutting systems are normally not applied to these species (Joyce 1998, Schütz 1997, Boudru 1986).

Single tree cutting

"Single tree cutting" is not strictly limited to the cutting of individual trees. The system also includes cutting very small groups, up to an area of some ten acres, corresponding to some ten mature trees.

Since common ash, sycamore and wild cherry are species that typically occur in intimate mixtures, it is normal that final cutting occurs at a single tree level. This means that trees are felled when they have reached individual maturity which is determined by their dimension, timber quality, position in the stand, future potentials, etc.. Thus, not all trees are felled at the same time. On the contrary, cutting of valuable broadleaves is spread over a long period according to silvicultural, economic, nature conservation and other reasons. The felling period of the valuable broadleaved species can even be extended over several decades, especially in the case of sycamore in oak stands. In stands where these tree species are extremely rare, each tree should be retained as long as possible as a valuable mixture component, irrespective of its quality or growth potential. Rieder (1998) also refers to the special value of retained individuals of sycamore in typical beech-sycamore stands.

Individual tree cutting is common in forests originating from coppice with standards and in the French "futaies claires". It is quite common in oak high forests mixed with common ash, sycamore and other hardwoods. In exceptional cases, in Denmark for example, selection forests comprised of oak, ash and maple occur, as indicated by Schütz (1997). Special attention should focus on the maintenance of the involved species. Fortunately these species are normally able to regenerate and survive, as they occur on very fertile sites and are relatively shade tolerant (Joyce 1998, Savill et al. 1997, Boudru 1986) or, in some circumstances, extremely shade tolerant (Van Miegroet and Lust 1972). Common ash, sycamore and, to a lesser extent, wild cherry are able to regenerate in small openings and survive for some years. The species benefit from or even require shelter in their juvenile stage. However, it is often recommended to open the canopy after 5-7 years of growth since, after this point, their light demand will rapidly increase. In cases where regeneration fails, they can be replaced by artificial plantations. In this respect Savill (1991) refers to the recommendation of some authorities to exclusively plant cherry in small groups where groups should not exceed approximately 20 trees.

Group cutting

It is clear that a group cutting system is applied in beech and oak stands with a typical group mixture and in stands where common ash, sycamore, wild cherry or other species are dominating and are group-mixed. The final cutting group size is flexible, but normally ranges from 0.1 to 0.5 hectare. A lengthy discussion about the precise group-size is not worthwhile provided that the main requirements are fulfilled: namely that the specific species is adapted to the site and that the regeneration technique is suitable.

The duration of the final cutting is also flexible. It depends on several factors of which the regeneration and development situation of the future stand are of the most important. Final cutting of all trees at one time is possible but certainly not common. The maintenance of immature valuable trees and the provision of shelter are two elements that strongly require a relatively long regeneration period. Periods of 10–20 years are certainly recommended, especially in cases where the area is not too small and the site quality increases the shade tolerance of the seedlings and saplings.

Joyce (1998) and Burschel and Huss (1987) give a good description of a group shelterwood system for ash. They consider it as the best approach to a successful natural regeneration.

This rather well known system, in principle applicable for almost all tree species and high forest types, shows the following interesting points:

- the starting area is rather small, even less than 0.1 hectare;
- the canopy is removed at three separate times; one third of the original stocking is removed at each intervention (in principle more interventions are also allowed);
- at each intervention the groups are enlarged to the desired or needed area, resulting in a final group with an area of (at least) 0.3 hectare;
- progress of the operation depends mainly on the light requirements of the species; for ash it may take only two to five years before the canopy should be opened, and five years until the canopy is fully removed; however, when the individuals are less light demanding, due to specific site conditions, the removal of the canopy may be much slower.

This vision is supported by Nussbaumer (1999) who also suggests small starting areas (0.1 hectare), a rapid extension of the groups and a rapid opening of the canopy. Depending on the size of the stand, Joyce (1998) recommends to lay out at each intervention new groups approximately 60 m apart from each other. In this way the groups will be connected to each other after two or three interventions. However, since the areas with valuable broadleaved species are in principle rather small, it will not be necessary to open several gaps at the same time as is normally the case in larger forests.

It must be taken into consideration that the rotation period of common ash, sycamore and wild cherry is much shorter, half as long, than that of oak and beech. This means that the second generation of small intermixed groups of these noble species will be surrounded by the remaining beeches and oaks. As a result, strong competition will exist between the large surrounding trees and the small young trees, which often leads to a strong depreciation of edge trees of both types. This is mainly the case with beech, therefore groups of valuable broadleaved species should not be too small and a suitable treatment, one which pays special attention to the edge trees of beech stands, should be carried out.

In any case, group mixtures are the best way to support or increase the share and quality of valuable broadleaves, especially in beech stands (Richter 2001,

Nussbaumer 1999, Joyce 1998, Burschel and Huss 1987). For wild cherry, small groups are strongly recommended by Boudru (1986) for these same reasons.

Small stand cutting

Over the last decades a tendency to establish small pure plantations of common ash, sycamore or wild cherry occurred for several reasons (Schrötter 2001, Joyce 1998, Bessières 1992, Boudru 1986, Lanier 1986, Thill 1979, 1970). The size of these plantations usually varies around one hectare. Large monocultures of these species are not established mainly because suitable sites are not available and because in nature these species are not characterised by pure stands.

The characteristics of the final cutting system applied in this forest type are very simple:

1. final cutting occurs at a scale of approximately one hectare;
2. all the trees are cut at the same time.

The second characteristic is not absolute; some trees can be retained as shelter for silvicultural, nature conservation or landscape reasons. The retained trees may be removed at a later date, in a short or over a longer time period. The majority of the stand, however, should be cut at the same time.

Such forest types might be quite attractive for private forest owners who might expect a significant financial return over a relative short term. An important prerequisite, however, is that sites are suitable for the site specific tree species. As sycamore is probably the most flexible tree species within this group, this species can be expected to yield the best results in large pure stands. Joyce (1998) confirms that the few known stands of pure sycamore high forests are very impressive, both in terms of growth performance and stem quality.

Despite this fact, the establishment of pure valuable broadleaved stands requires due attention. Afforestation of former agricultural land meets the light demand of these species but presents shelter problems with respect to frost susceptibility. Pure ash planting should be avoided in areas with an increased frost risk. Furthermore, large monocultures tend to be more prone to insect attack by ash bud moth and to wind which can break the leader shoots (Joyce 1998). The success of sycamore is mainly due to its adaptability to a wide range of conditions from lowland to upland and its relative tolerance to frost and exposure. Wild cherry, as a tree species of the subatlantic-submediterranean climatic region, shows a marked preference for warm and sunny sites and is susceptible to damage by late spring frosts. These plantations therefore require a side shelter (Boudru 1986)

4.4.3 Rotation period

The rotation period is a major characteristic of the final cutting system. However, as explained above, due to several reasons a fixed rotation period cannot be considered. Thus the rotation period should be considered as an indicative factor of stand development.

With respect to the involved valuable broadleaves, an interesting point can be noticed: the rotation is similar for all species. Generally speaking most authors suggest a rotation period of 60–80 years (Schrötter 2001, Joyce 1998, Boudru 1986, Thill 1975, 1970). Of course there exist several circumstances where other rotations are applicable. Cherry plantations can already be harvested with a rotation of 50 years, especially on very fertile sites and in cases where the aim is not to achieve large dimensions. In contrast, it is quite normal to have sycamore rotations equal to 100 years (Rieder 1998).

The rotation period is determined by several factors of which site characteristics, growth potentials, production goals and timber quality are major elements.

Site and growth potentials

All valuable broadleaves are very site-specific (Joyce 1998, Savill 1991, Boudru 1986). If they are not planted under optimal conditions their growth potential declines and their susceptibility to diseases increases.

The short rotation of valuable broadleaves is partly determined by the natural growth pattern of these species. They experience fast growth in their youth and growth culminates very early. On suitable sites they show excellent growth performances over the first thirty years, but after this period growth tends to decline and beyond 50 years it falls off considerably (Lüdemann 2001, Joyce 1998, Faust 1963) However, yield models of common ash, sycamore and wild cherry strongly differ according to the authors. For instance, the French yield model for common ash shows rapid height growth with a dominant height of 30–34 m at 80 years (Franc and Ruchaud 1996, Lanier 1986), whereas the British model for common ash, sycamore and birch for the yield class 10 shows a top height of 24.6 m (Joyce 1998, Hamilton and Christie 1971). The Dutch best yield class 9 mentions a dominant height of 30.3 m at an age of 80 years for this species (Schütz and Van Tol 1981).

Stand treatment also strongly affects growth pattern and consequently rotation period. The involved species also have similar features in this respect. Although they are for the most part quite shade tolerant in their youth, they later become strong light demanders which require sufficient light to grow satisfactorily. From as early as the thicket stage, they require adequate growing space. Heavy thinnings are needed throughout their life to attain the desired rate of diameter growth. Because of this, it became usual in the last decades to select a certain number of final crop trees from an early stage. Final crop trees of common ash, sycamore and wild cherry are selected at top heights of some 15 m and after successive thinnings have already occurred (Nussbaumer 1999, Joyce 1998, Lanier 1986).

Production goals

It is evident that all valuable broadleaves have the same major production goal: high quality timber production, meaning straight, defect- and branch-free boles with a minimum length of 6–7 m and a d.b.h. of 50–60 cm. However, this objective is rarely achieved as indicated by Foet (2001).

It is quite clear that rotation time is less important when no valuable timber can be produced. In cases such as this, Frank (1973) suggests a rotation period as short as 30 years for simple mass-production of sycamore in Denmark.

Some differences between the species should be underlined. The primary objective with ash should be the production of veneer logs for the furniture industry and sawtimber. High quality timber logs should be 50–60 cm at d.b.h.. In addition there is a specific goal, mainly appreciated in United Kingdom, which focuses on the production of material for the manufacture of hurleys. This kind of material should be 28–32 cm in diameter for optimal conversion. In a study conducted by the Forest Service it is assumed that hurleys can be produced on a 25 years rotation, although Joyce (1998) believes that a rotation period of 40 years is more realistic. The objective for sycamore, like common ash, should also be the production of logs for veneer and sawtimber. Veneer material requires boles of 40 to 60 cm at d.b.h.. High priced stems of wild cherry must have a mean mid diameter greater than 45 cm with a target diameter of 60 cm. However, log sizes down to 24 cm top diameter are already acceptable by the furniture trade in Britain. Planking logs require a mid-diameter greater than 35 cm (Schrötter 2001, Joyce 1998).

Timber quality

Some factors of timber quality directly affect the rotation period such as the desired width of the annual rings and the eventual deterioration of the wood after some age or diameter has been reached.

Regular and large annual rings are necessary in common ash to produce good quality wood. A ring width of 4–5 mm (0.8–1 cm annual diameter growth) is required for quality sawlog material. Smaller ring widths yield wood of lesser strength which can be used for veneer, while larger ring widths yield better mechanical properties that are favourable for products which require greater strength, flexibility and elasticity (Nussbaumer 1999, Joyce 1998). Annual rings of wild cherry tend to vary over a wide range. Thill (1975) and Boudru (1986) prefer an annual circumference increment of 2.0–2.5 cm.

A specific and extremely important factor determining the rotation is the great risk of timber discoloration and deterioration after the age of 60–80 years. Very well known is the black or brown heart which usually appears in common ash at the age of 60 years and with circumferences of more than 120–150 cm (Nussbaumer 1999, Boudru 1986, Lanier 1986). Less important, although noticeable, is the brown heart in sycamore which also appears after the age of 60 years (Foet 2001, Rieder 1998, Thill and Mathy 1980). In rare cases even heart rot may occur. Wild cherry is susceptible to heart rot can easily become severely deteriorated (Thill and Mathy 1980). Foet (2001) suggests aiming for stems with a diameter of 45-60 cm since he has noticed that discoloration begins after 60 cm.

4.4.4 Final cutting of additional valuable broadleaves

As the above review reveals, information and knowledge pertaining to final cutting systems of the most common valuable valuable broadleaves is still poor. It is clear that information about final cutting systems of additional valuable broadleaves is even scarcer (Heuer and Jablko 1998). Only of Norway maple (*Acer platanoïdes L.*) and of wild service tree (*Sorbus torminalis L.*) could some primary information be collected. Suitable data in the literature on walnut *(Juglans regia L.*), black walnut (*Juglans nigra L.*), crab apple (*Malus sylvestris Miller*), whitebeam (*Sorbus aria (L.) Crantz*) and other hardwoods are extremely scarce, if not nonexistent (Foet 2001, Kristöfel 1998, Savill 1991).

It appears, however, that there are no major differences between species like Norway maple and wild service tree and between common ash, sycamore or wild cherry: These broadleaves mainly survive in coppice and coppice with standards and rarely appear in high forests. Under such circumstances it is not as significant to discuss final cutting systems. Rarely occurring trees must be protected and retained as monuments and seed trees as long as possible.

Schrötter (2001) mentions that Norway maple occurs in Mecklenburg-Vorpommern. From existing experience it was concluded that the maintenance of this rare species should be encouraged by managing it in groups or in small stands (Schrötter 2001, Heuer and Jablko 1998). It occurs in stands mixed with sycamore within an area of 0.55 ha on average. A few stands are larger than one hectare and only two are larger than three hectares. Savill (1991) points out that Norway maple should be planted in groups rather than in intimate mixtures because its rapid early growth may cause suppression of the other mixed species.

Final cutting systems for Norway maple will therefore not differ significantly from the systems used for common ash, sycamore and wild cherry. Individual tree cutting should be carried out in intimate mixtures and group cutting or small stand cutting in larger plantations. Rotation for Norway maple might be longer than for sycamore, according to Schrötter (2001) who proposes a production time of 100–120 years and an average diameter for logs of more than 40 cm under bark. This is in contrast to the findings by Heuer and Jablko (1998) who did not find growth differences between sycamore and Norway maple. Therefore rotations of 80 years should be possible.

Schrötter (2001) also mentions that wild service tree is presently found in remnants of coppice or coppice with standards stands. For this species, small scale plantations are preferred to intimate mixtures. He suggests plantation of wild service tree on areas of 0.5-1.0 ha for the production of poles, with a minimum average diameter of more than 40 cm, in a period of 120-150 years.

References

Bessières, F. 1992. La conduite des peuplements de frêne (*Fraxinus excelsior* L.) et de mérisier (*Prunus avium* L.). Revue forestière française. Pp. 115–120.

Boudru, M. 1986. Forêt et sylviculture – Sylviculture appliquée. Les Presses agronomiques de Gembloux. 244 p.

Burschel, P. and Huss, J. 1987. Grundriß des Waldbaus. Verlag Paul Parey, Hamburg und Berlin. 352 p.
Christensen, M. and Emborg, J. 1996. Biodiversity in natural versus managed forest in Denmark. Forest ecology and management 85: 47–51.
Dengler, A. 1971. Waldbau auf ökologischer Grundlage. Verlag Paul Parey, Hamburg und Berlin. 229 p.
Evans, J. 2001. The Forests Handbook Volume 2. Blackwell Science, London, Edinburgh. 382 p.
Faust, H. 1963. Waldbauliche Untersuchungen am Bergahorn. Dissertation Göttingen. 146 p.
Foet, C. 2001. Zielstärkennutzung von Edellaubholz im Forstamt Hadamar. AFZ-Der Wald 13: 669–671.
Franc, A., Bolchert, C. and Marzolf, G. 1992. Les exigences stationelles du Mérisier: revue bibliographique. Revue forestière française. Pp. 20–26.
Franc, A. and Ruchaud, F. 1996. Autécologie et feuillus précieux: frêne commun, mérisier, érable sycomore, érable plane. Cemagref. 170 p.
Frank, H.J. 1973. De bergesdoorn in de Nederlandse bosbouw. Scriptie, Wageningen. 83 p.
Hamilton, G.J. and Christie, J.M. 1971. Forest management Tables (Metric). Forestry Commission, Booklet no. 14. London, Her Majesty's Stationry Office. 201 p.
Heuer, L. and Jablko 1998. Spitzahorn ist eine forstlich interessante Baumart. AFZ-Der Wald 18: 956–958.
Houllier, F. and Rittié, D. 1992. Eléments sur la ressource en feuillus précieux. Revue forestière française. Pp. 13–19.
Joyce, M.J. 1998. Growing broadleaves. COFORD, National University of Ireland, Belfield, Dublin 4. 144 p.
Kazda, M. and Pichler, M. 1998. Priority assessment for conversion of Norway spruce forests through introduction of broadleaf species. Forest Ecology and Management 102 (2–3): 245–258.
Kristöfel, F. 1998. Zum Wachstum von Juglans nigra in den Leithaauen in Ostösterreich. Forst und Holz 53: 43–47.
Lanier, L. 1986. Précis de sylviculture. Nancy, ENGREF. 468 p.
Lévy, G., Le Goff, N., Lefèvre, Y. and Garros, L. 1992. Les exigences stationelles du frêne dans le Nord et le Nord-Est de la France. Revue forestière française. Pp. 20–26.
Lüdemann, G. 2001. Von der Pappel zum Edellaubholz. Forst und Hol 56 (3): 67–68.
Matthews, J. D. 1989. Silvicultural Systems. Clarendon Press, Oxford. 284 p.
Mayer, H. 1977. Waldbau auf soziologisch-ökologischer Grundlage. Gustav Fischer Verlag, Stuttgart, New York. 483 p.
Neirynck, J., Mirtcheva, S., Sioen, G. and Lust, N. 2000. Impact of *Tilia platyphyllos* Scop., *Fraxinus excelsior* L., *Acer pseudoplatanus* L., *Quercus robur* L. and *Fagus sylvatica* L. on earthworm biomass and physico-chemical properties of loamy topsoil. Forest Ecology and Management 133(3): 275–286.
Nussbaumer, H. 1999. Buchen, Eichen und Eschen in den Wäldern am Bodensee zwischen Arbon und Ermatingen. Schweiz. Z. Forstw. 150(7): 252–256.
Revue Forestière Française 1992. Commercially valuable occasional hardwoods: ash, wild cherry and maples. 188 p.
Richter, J. 2001. Eichen-Edellaubholz-Bestände: an der Schwelle zum Dauerwald. Forst und Holz 56 (1): 9–11.
Rieder, A. 1998. Ahorn-Wertholzproduktion in kurzen Umtrieben. AFZ-Der Wald 15: 776–779.
Savill, P. 1991. The Silviculture of Trees used in British Forestry. C.A.B. International, U.K. 143 p.

Savill, P., Evans, J., Auclair, D., and Falck, J. 1997. Plantation Silviculture in Europe. Oxford university press, Oxford, New York Tokyo. 297 p.
Schrötter, H. 2001. Vogelkirsche, Spitzahorn und Elsbeere: Drei wertvolle Baumarten in Mecklenburg-Vorpommern in Abseits. Forst und Holz 56: 188–196.
Schütz, J.P. 1997. Sylviculture 2. Presses polytechniques et universitaires romandes. CH-1015 Lausanne. 178 p.
Schütz, P.R. and Van Tol, G. 1981. Aanleg en beheer van bos en beplantingen. Wageningen, Centrum voor landbouwpublikaties en landbouwdocumentatie. 504 p.
Thill, A. 1970. Le frêne et sa culture. Gembloux, Presses Agronomiques de Gembloux. 85 p.
Thill, A. 1975. Contribution à l'étude du frêne, de l'érable sycomore et du mérisier. Bull. Soc. R. For. Belg. (1): 1–12.
Thill, A. 1979. La sylviculture du frêne en Belgique. Bull. Soc. R. For. Belg. 86(3): 124–129.
Thill, A. and Mathy 1980. La culture des essences précieuses en Belgique. Ann. Gembloux 86 (1): 1–32.
Van Miegroet, M. and Lust, N. 1972. Aufbau, Wachstum und Reaktionsvermögen von unterdrückten Eschenverjüngungen. Sylva Gandavensis 34: 1–38.

4.5. Wood Properties and Wood Processing of Valuable Broadleaved Trees Demonstrated with Common Ash and Maple in Southwest Germany

Gero Becker and Joachim Klädtke
Institute of Forest Utilization and Work Science, University of Freiburg, Germany

Abstract

Among the broadleaved tree species in Central Europe, common ash *(Fraxinus excelsior [L.])* and maple, especially sycamore *(Acer pseudoplatanus [L.])*, are very important in terms of value production. Due to their specific visual/aesthetic and mechanical properties as well as good tooling ability, they are an excellent material, as solid or glued, for all kinds of furniture and indoor construction, whereas outdoor and structural usability is limited because of the low natural durability of the wood. Logs of larger dimensions are used to produce sliced veneer as a decorative surface for furniture and flooring. Despite the generally good mechanical properties, the uses of ash and maple are few in load-bearing structures. Silvicultural concepts aiming to achieve superior quality and big dimensions in a relatively short time by heavy crown thinnings are to be favoured.

Keywords: common ash, Fraxinus excelsior [L.], maple, sycamore, Acer pseudoplatanus [L.], wood quality, wood properties

4.5.1 Introduction

Although *Fraxinus excelsior* (common ash) and *Acer* (maple) are subordinated in Central Europe to beech and oak in terms of forest area, standing volumes and annual cut, they are of high economic importance due to their excellent wood properties and their suitability for top quality furniture and joinery production.

Figure 4.29 gives some information on the annual cut of logs of common ash and maple in comparison to beech *(Fagus sylvatica)* for the area of Baden-Württemberg (LFV BW 1989–2002).

Whereas the annual cut of beech over the past decade reaches in its peak up to 900,000 m^3y^{-1}, the trend is nearly constant for common ash with about 25,000–30,000 m^3y^{-1} and shows a slightly increasing tendency for maple from 1989 to 2002.

In addition, Figure 4.30 shows the share of the best and 2nd best log grades in the total harvested volume for the three tree species. The portion of the best log grade A/TF/T is far below 10% for beech and maple, and only marginal for common ash.

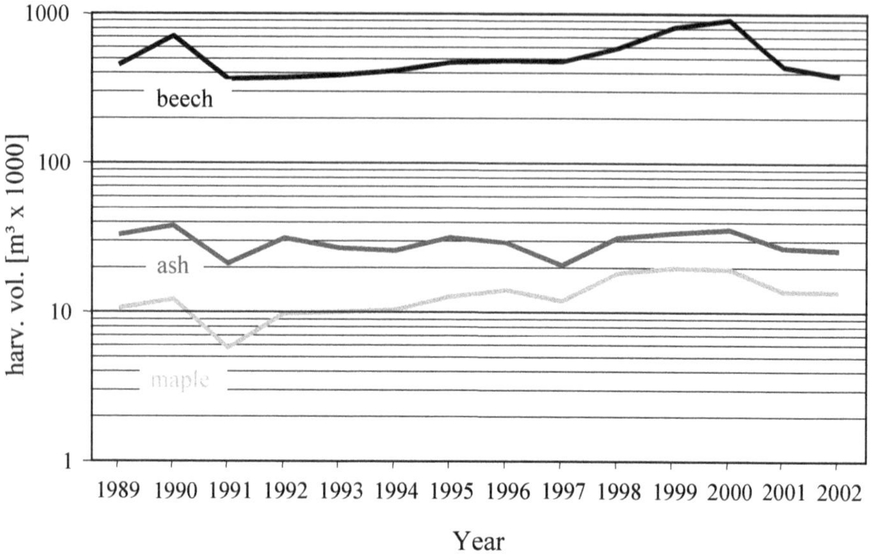

Figure 4.29. Annual cut of beech, ash, and maple in Baden-Württemberg.

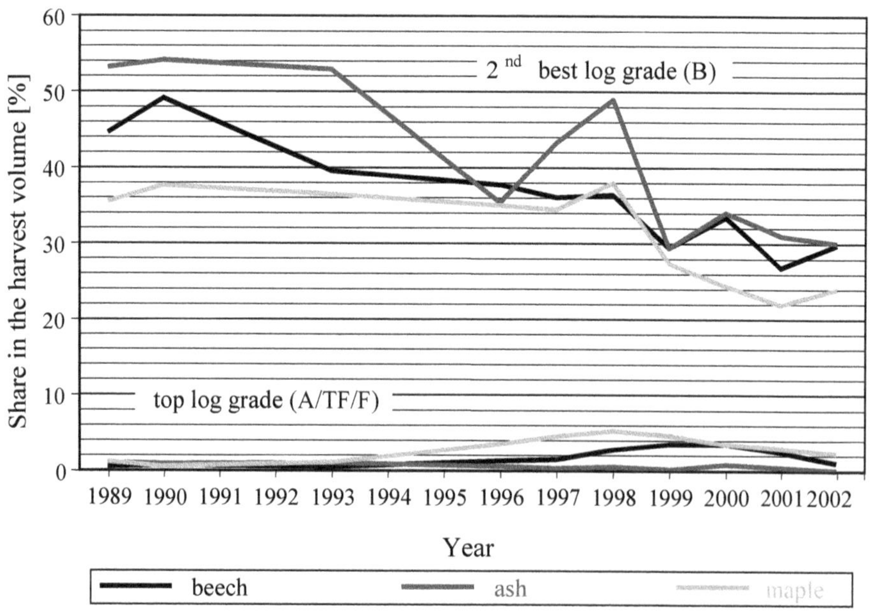

Figure 4.30. Share of the best and 2nd best log grades in the total harvested volume.

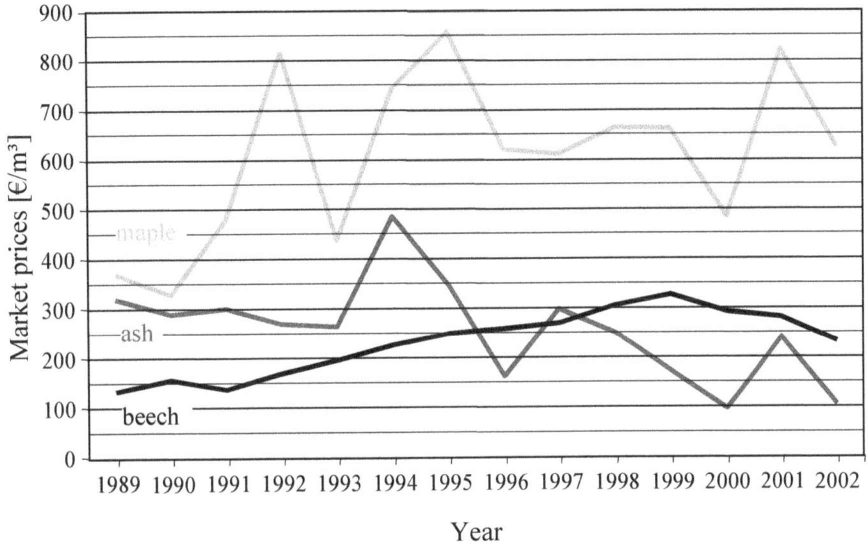

Figure 4.31. Prices per m³ for maple, ash and beech (best log grade A/TF/F).

The trend of the second best log grade is quite similar for all the three tree species, decreasing from the early 1990s to only 20–30% in 2002.

Figure 4.31 (LFV BW 1989–2002) shows the development of the market prices for logs of the highest quality class F/TF/A of common ash, maple and beech over the past decade. The significantly and steadily increasing market value of beech during the last decade is opposite to decreasing prices for ash, which is currently within the range of beech. Top quality logs of maple are much better paid than those of common ash and beech, although the market prices fluctuate considerably.

4.5.2 Common ash

Wood characteristics and technological properties

Fraxinus is a ring-porous hardwood. The early wood vessels show wide lumina, the late wood vessels have thicker walls and, as a consequence, the wood has a clear and distinctive structure of year rings, with a light yellow colour throughout the cross section. At higher ages, starting at approx. 60 years, *Fraxinus* can develop facultative heartwood with a darker (red-brown-green) colour. The discoloration of the heartwood has no technological consequences but is not desired by the customers in furniture and joinery branches and, therefore, results in a severe reduction of the price.

The wood of *Fraxinus* has a high density ($\rho_{15} \sim 0.69$ gr/cm³), which is at similar level to beech and oak. In line with the high wood density, the technological properties are excellent (Table 4.20, LWF 2002), where the values given in Table

Table 4.20. Stiffness and strength properties of clear wood of ash and maple compared to beech and oak (LWF 2002).

	modulus of elasticity [Nmm^{-2}]	tensile strength [Nmm^{-2}]	compr. strength [Nmm^{-2}]	bend. strength [Nmm^{-2}]	max. load [kJ/m^2]
Maple	9400–11400	82–144	49–58	95–112	62–65
Ash	13000–13400	130	44–52	102–120	68
Beech	14000–16000	135	53–62	105–123	100
Oak	11700–13000	90–110	52–61	88–95	60–75

4.20 correspond with the *average* values published by Wagenführ (1996). As it is typical for ring porous hardwoods, the mechanical property improves with increasing ring width. This means that the faster the growth, the better the strength and stiffness properties of *Fraxinus*.

The dimensional stability of common ash wood is about similar to oak but better than for beech. The tooling ability of common ash wood is excellent, and, after planing and sanding, it shows a desirable clear and smooth surface.

Due to the lack of extractives, common ash wood has a low natural durability. Because ash wood is difficult to impregnate, outdoor use of ash wood is not recommended.

Figure 4.32 (LFV BW 1989–2002) shows the relationship of the market price per m³ of common ash by log grade. The prices for the best grades fluctuate within a broad range, but generally with a decreasing tendency since 1989). In contrast, the actual prices for the second quality B are only about 50% of the prices in 1989. The prices for the inferior qualities have been more or less stable and do not differ much between quality C and D.

Utilisation

Due to the light, neutral colour and the excellent mechanical properties, common ash wood is a good material for all kind of massive high performance indoor use (furniture, stair cases, etc.). Large dimension logs (mid diameter >30 cm) can be used to produce sliced veneer as decoration surfaces for doors, furniture, etc. Laminated common ash veneers make high performance boards. Because common ash wood is very hard, it is excellent for flooring (parquet). In addition, due to its high strength and stiffness, it is, together with hickory, the preferred wood choice for tool handles (e.g. for axes) and sports equipment (barren, skies, ladders).

In former times, the excellent combination of static and dynamic mechanical properties and good tooling ability made *Fraxinus* the preferred material for carriages, railway cars and even motorcars of the early years. It was also used for specific parts of textile and agricultural machinery. *Fraxinus* wood with wide rings (e.g. wood of young and/or fast grown trees) is still preferred for a spectrum of special uses. However, despite the good mechanical performance, common ash is hardly used in load-bearing structures (Begemann 1981–1994).

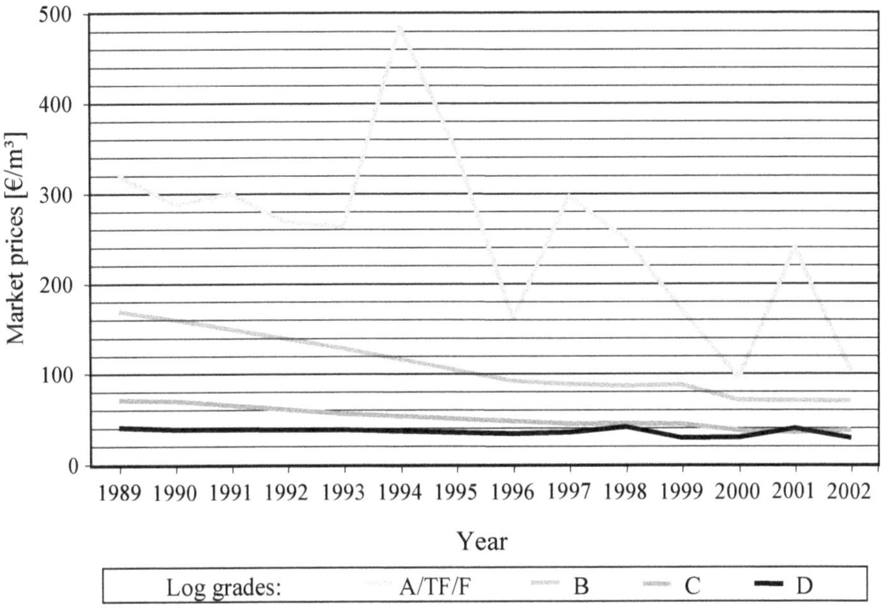

Figure 4.32. Market prices of ash 1989–2002 for different log grades.

4.5.3 Maple

Wood characteristics and technical properties

Among the different maple species in Germany, only *A. pseudoplatanus* and *A. platanoides* are relevant for forestry and the wood market.

Maple is a diffuse-porous hardwood species. The size of the vessels does not differ much between early and late wood, although the border between the year rings is visible because the last row of late wood cells is of a slightly darker colour. The colour of maple is between very light yellow and white. The colour is uniform throughout the cross section, and there is no clear boarder between hardwood and sapwood.

Some maple logs show a facultative heartwood-discoloration (grey-dark-brown) which leads to a significantly lower price. Because the risk for this discoloration increases with the increasing age of the tree, heavy thinnings from above are recommended for maple alike to all valuable broadleaved tree species (Hein 2004).

For maple, the density ρ_{15} is approximately 0.63 gr/m³. The static and dynamic strength properties are excellent, but stiffness does not reach the high values of *Fraxinus*. The dimensional stability is excellent and beats beech, oak and ash. Maple has a low natural durability, but due to its diffuse-porous wood structure it is easy to impregnate.

The excellent dimensional stability and the homogeneous light colour are the most appreciated properties of *Acer* species. During processing and drying, extreme conditions should be avoided; otherwise the light colour may turn into dark yellow or even brown.

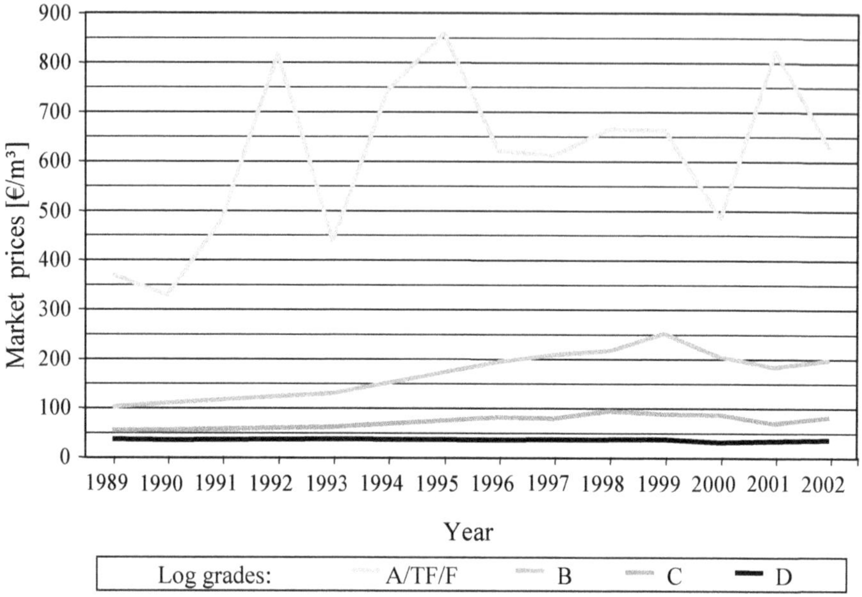

Figure 4.33. Market prices of maple 1989–2002 for different log grades

Figure 4.33 (LFV BW 1989–2002) shows the relationship of the market price per m³ by log grade. It is obvious that only the quality classes F/TF/A and B are valuable. For the top quality class F/TF/A, a mid diameter of at least 30 cm is needed, and no severe defects, knots, spiral grain, etc., are allowed (MLR BW 1983).

Utilisation

The wood of *Acer* is an excellent material for all kinds of furniture, especially tables, and for indoor constructions (stairs, cabinets). Due to its hardness it is also appreciated for flooring and parquet. Typical utilisations are tables in pubs and traditional restaurants.

As in the case of common ash, logs of larger dimensions (> 30 cm mid diameter) are used to produce sliced veneer as a decorative surface for furniture and flooring.

The dimensional stability also makes maple suitable for carving and turning. However, despite these desirable properties, maple is scarcely used in load-bearing structures (Begeman 1981–1994).

4.5.4 Conclusion

Due to their excellent suitability for top class furniture and indoor usage, common ash and especially maple are very important broadleaved tree species in Central Europe in terms of value production. However, this counts only for the best wood qualities, requiring faultless properties and log diameters of at least 30 cm. Inferior qualities, devaluated by knots, unfavourable stem form, etc., and smaller dimensions hardly achieve cost covering market prices. Unfortunately, the proportion of better qualities has been insufficient until today, and needs to be necessarily increased. Therefore, silvicultural concepts which aim to achieve superior quality and big dimensions in a relatively short time by heavy crown thinning are to be favoured.

References

Begemann, H. 1981. Das große Lexikon der Nutzhölzer, vol. 1 & 2. Dt. Betriebswirte-Verlag, Gernsbach.

Hein, S. 2004. Zur Steuerung von Astreinigung und Dickenwachstum bei Esche (*Fraxinus excelsior* L.) und Ahorn (*Acer pseudoplatanus* L.). Schriftenreihe Freiburger Forstliche Forschung; Bd. 25.

LWF 2002. Beiträge zur Esche. Berichte aus der Bayerischen Landesanstalt für Wald und Forstwirtschaft. 89 p.

LFV BW 1989–2002. Forststatistische Jahrbücher. Landesforstverwaltung Baden-Württemberg, Eigenverlag.

MLR BW 1983. Die Handelsklassensortierung für Rohholz mit ergänzenden Erläuterungen (Forst-HKS). Ministerium für Ländlichen Raum, Ernährung, Landwirtschaft und Forsten Baden-Württemberg. 24 p.

Wagenführ, R. 1996. Holzatlas, 4. Auflage. Leipzig, Fachbuchverlag. 688 p.

5. Environment and Society

5.1 Valuable Broadleaved Trees in the Landscape

Simon Bell
United Kingdom

Abstract

Valuable broadleaved trees form a component of the landscape of many parts of Europe, in forests, in the wider rural landscape and in urban areas. Aspects of forest landscape aesthetics, now a well developed discipline, apply to valuable broadleaved trees as much as to other trees. Across Europe there is a range of different "forest cultures" which affect the way people relate to forests and trees, so that any consideration of forests and landscapes has to be viewed in that context. Valuable broadleaved trees such as ash, cherry and sycamore contribute to the forest landscape in terms of species patterns expressed as colours, textures and seasonal variation. In the rural landscape they have a special role to play, especially in the case of cherry. In the urban scene ash and sycamore are less commonly used, while cherry is very popular, although frequently as non-native species or special ornamental varieties. When grown intensively for higher timber volume and quality the same design principles should be applied as for other forest types. Plantation layout, shape and scale of woodlands and compartments, the geometry of rows, stand structure and the effects of final cutting all need consideration.

Keywords: forest landscape design, landscape aesthetics, landscape perception

5.1.1 Introduction

At first sight it may seem that a lengthy consideration of the place of valuable broadleaved trees such as ash (*Fraxinus excelsior*), cherry (*Prunus avium, P. padus*) and sycamore (*Acer pseudoplatanus*) in the landscape is unnecessary. After all, these are trees native to much of Europe and occur as natural stand components of many forests. They contribute colour and texture to the landscape. They are also commonly found in city parks and streets and in the wider countryside.

This attitude takes the presence of such trees as positive elements in the landscape for granted and, when they are valuable for timber, ignores the possible negative aesthetic effects of intensive management. Aesthetic values are increasingly becoming recognised as more than a cosmetic concern of superficial appearances. Landscape values may also be measured in economic terms, enabling more informed judgements to be made about the balances between timber and other values.

This chapter covers two main areas. Firstly there is the contribution of valuable broadleaved trees to the landscape as a whole and secondly the implications for the landscape of intensive silvicultural management aimed at increasing volume and quality of valuable broadleaved timber. Before covering the detailed aspects of this subject it is also worth reviewing some of the main elements of the theory and practice of forest landscape aesthetics, a subject that has received significant attention over the last couple of decades. As valuable broadleaved trees are not yet a common element in the landscape, few if any studies of their landscape and visual impact have been undertaken. Therefore this chapter is more of a speculation of the kinds of issues likely to come to the fore and the ways these might be dealt with. The application of design has been made from the point of view of an expert professional designer extending the use of the basic principles into this field; an approach that has been successful in other areas of woodland design involving the use of unusual elements such as agro-forestry or short rotation coppice.

Recent research has synthesised the different approaches of the fields of aesthetic philosophy, environmental psychology and design in order to gain an understanding of the deeper significance of forest aesthetics. Europe is not a homogeneous area, either socially or ecologically, and it is as impossible to apply generalized concepts of landscape aesthetics across the continent as it is with forest management models. Thus, some understanding of the different "forest cultures" of Europe is essential. Research has also been carried out into public perceptions of forest and countryside landscapes, the results of which may differ from the views of foresters. Results are unavailable for the whole of Europe, but the main findings of those studies that have been carried out will be briefly reviewed because they have implications for the use of valuable broadleaved trees.

The contribution of valuable broadleaved trees to the landscape will be classified into three types:

1. As components of larger scale forest and woodland landscapes where species patterns, textures, colours and seasonal variation as well as the qualities of different stand types, edges and small-scale groups contribute to the forest aesthetic. The ecological and silvicultural contributions of these trees have to be balanced with their aesthetic role in the larger forest.
2. As part of rural landscapes where small woods and copses, trees in hedgerows, fields and meadows, as well as near farmhouses and along roadsides, where they provide landscape structure and diversity. In this type of area the contribution of these trees to the cultural as well as aesthetic landscape has to be considered.
3. As part of the urban landscape, where urban woodland, park and street trees and trees in gardens give contrast to the built environment. The recreational and aesthetic roles of these trees are interrelated in this situation.

The implications for the landscape of intensive silvicultural management will include the application of some basic forest design principles that have been successfully applied over several decades. Some specific design aspects of plantation layout, such as the shape and scale of woodlands and compartments, the geometry of rows of trees, stand structure, final cuttings, species mixtures and non-clearfell systems, will also be applied.

5.1.2 What is a landscape?

In a brief overview such as this chapter, it is not possible to present a deep analysis of the different meanings that may be applied to the term "landscape". The word landscape, as it is used in the English language, has a range of applications. The term is often considered synonymous with aesthetic values. In this section the meaning of the two terms will briefly be reviewed in order to establish context for the discussion that follows.

The term landscape can be used in a variety of ways: as a scale of planning, as an approach to forest ecology and as the perceptual realm of the senses. The word itself originates in an old German word "lantscaf". "Scaf" became the English word "shape" and the German "schaben" or "schaffen", so that, for example, the current German "Landschaft" means land that has been shaped (by man) (Haber 2002). In English the influence of Dutch landscape painting on British art and landscape appreciation has led to the development of the dictionary definition of "landscape" as a "prospect of scenery that can be taken in at a glance from one point of view" (Oxford English Dictionary). Thus, landscape is both a physical and an experiential concept.

The application of the idea of landscape to forests and woods varies due to the three-dimensional properties of the height of trees and their ability to enclose space:

"Woods are not like other spaces. To begin with, they are cubic. Their trees surround you, press you in from all sides. Woods choke off views and leave you muddled and without bearings. They make you feel like a small child lost in a crowd of legs." Bill Bryson: A Walk in the Woods (Bryson 1998)

As is evident from this quotation, woodlands are places where we experience our surroundings differently from the countryside in general or the urban scene. We are used to thinking of the landscape as a scene that we observe, from a viewpoint or as we pass by in a car or train, using mainly our eyes for that observation. When we are in the woods, however, we use all our senses to the fullest extent. These senses include not only sight, hearing, taste, smell and touch but also information from balance, from a sense of going up or down, side-to-side and twisting and turning. Our skin is sensitive to temperature and humidity. These are called the haptic or kinaesthetic senses (see below).

The nature of forest aesthetics

Aesthetics refers to things perceived by our senses and the reactions we make to such perceptions (Bell 1999). Forests are particularly important elements in any landscape because of the way they tend to affect what and how much we perceive It is useful to consider two related aspects of forestry aesthetics. These are the traditional, scenic mode of viewing and the participatory, or multi-sensory mode.

The scenic mode represents the typical static viewpoint offering a prospect of scenery, the traditional definition as quoted above. The forest in this aesthetic mode acts as a component of the vista; it is an external view over the canopy of the forest. We perceive the landscape primarily visually. We comprehend patterns made up of shapes, colours, textures or lines and their different arrangements. In a natural forest, unaffected by deliberate human activity such as logging (but possibly

Figure 5.1. In this Bavarian scene, the typical scenic mode of viewing is experienced. The scene fits the concept of "beauty" in a picturesque sense. Photo: Simon Bell.

affected indirectly through fire or disease control), these patterns arise from the interaction of ecological processes with climate and landform.

In landscapes changed by human activity, the patterns remain influenced to greater or lesser degrees by natural processes but also include a range of human caused patterns. These include logging, clearing for agriculture, plantation forestry and sundry associated activities such as road construction. Some of these patterns may have long histories, such as those of small woodlands linked by hedges in England or France (Hoskins 1995, Rackham 1986), and may never have been planned for aesthetic values, although many people find them very attractive.

The aesthetic experience of the landscape scenery tends to be one of two varieties – a sense of beauty or of the sublime (Bell 1999, Burke 1958). The sense of beauty is a positive emotion which is induced when we find the patterns we see to be arranged in such a way that they fit together harmoniously and appear to be of one piece with no joining elements. The sublime is experienced when we are faced with limitless expanses of a scale and magnitude that dwarfs us, and when the forces of nature seem to overwhelm us. This is not necessarily a positive emotion but it can nevertheless be a powerful one.

A good deal of work has been carried out over the years, mainly since the 1960s, attempting, quite successfully, to find out what kind of forest landscape people prefer (Brush and Shafer 1975). This preference research often uses photographs of different scenes and asks members of the public to evaluate them by expressing which they liked most or least, for example, and why that might be. It can be argued that most people tend to prefer certain kinds of scenery, although the extent to which this is cross-cultural is not clear. Two of the main researchers in this field, Americans Steven and Rachel Kaplan, have identified some key factors, which they believe account for much of the expressed preferences. These factors are not objectively measurable attributes of the scenes under evaluation, but are cognitive factors; that is, they are concerned with how we make sense of and interpret the scene (Kaplan et al. 1998). These factors are as follows:

Figure 5.2. Walking along a path through the forest permits the participatory mode of landscape aesthetics to take place. We can use the range of our senses to gain a rich experience of the landscape. Photo: Simon Bell

- **Coherence**, where the scene is comprehensible, the patterns are understood and the structure is clear.
- **Legibility,** where the elements of the landscape are so arranged as to enable us to read it and make sense of it.
- **Complexity**, where there is an abundance of variety, where the structure is not simple.
- **Mystery**, where parts of the scene are hidden from direct view, where it has to be explored, where not all of it is visible at once.

The second mode of aesthetic interaction is the participatory aesthetic, or the aesthetic of engagement (Berleant 1982). This interaction makes use of all our senses (as mentioned above in relation to the Bryson quote). It is participatory because we are physically in the landscape as opposed to seeing it from a distant viewpoint. This is where the forest encloses us and can shut us off from external surroundings. Such close proximity to the forest, experienced perhaps from a path, can be very rewarding. We can be assailed by many stimuli: the sight of the trees, shrubs, flowers, rocks, wildlife; the sound of the wind in the trees, of bubbling water, of birdsong; the smell of hot tree resin, flowers, fresh soil; the tactile surface of bark or rock, the coolness and dankness of deep shade and the warmth and dryness of bright sun. In a forest all these stimuli work together harmoniously (in comparison with many urban landscapes) and induce the sense of beauty described above.

Depending on the topography, some places tend more towards the scenic mode, others to the participatory mode of aesthetic experience. In mountainous or hilly countries, where unforested mountain tops or cleared valley bottoms enable large-scale distant views to be obtained, the scenic mode is often dominant. Forested hills or mountainsides may surround settlements; scenic routes used by tourists may exploit these areas and both the beauty and sense of sublime make them prime candidates for national parks or other designations of protected status. It is also possible to experience the participatory aesthetic, yet for some reason the open vistas and dramatic views are far more significant. The Swiss and Austrian Alps, the Pyrenees, Dolomites or the mountains of Norway, as well as the smaller ranges of the Scottish Highlands, the Tatras in Poland or Halkidiki in Greece and so on, are all examples of splendid forest scenery.

Countries with less mountainous topography often offer a completely different experience and the relationship of the inhabitants to the forest is often more direct and integral to their everyday lives. Finland is dominated by forests, and all the farms appear to occupy spaces carved out of them. The same applies to large parts of Sweden, Russia west of the Urals and some of the Baltic countries.

Thus, the consideration of forest aesthetics must deal with both modes of experience. In the past, aesthetic concerns were raised by the public over, for example, the extensive conifer plantation re-afforestation on the open hillsides of Britain. Thus the major developments in forest aesthetics have tended to concentrate on scenery management and designing the appearance of the forest in the wider landscape (Bell 1999, Lucas 1991, Forestry Commission 1989). Only recently has significant attention focused on the internal design of flatter areas where the participatory aesthetic tends to dominate.

Regional differences in forest culture across Europe

When trying to synthesise aspects of forest landscape aesthetics and design across Europe, it is important to ensure that regional differences are highlighted. It is clearly the case that the character and function of forests, rural landscapes, urban parks and street trees contain both similarities and differences between, for example, England, Austria, Italy and Finland. People undertake popular activities such as dog walking, jogging, or picnicking in all four countries. However, the climate, ecosystem, cultural history and character of the people also major differences; therefore it would be inappropriate and undesirable to recommend principles of design that ignore this rich variety.

Between the European-wide and the national or local scales it is possible to detect broad regional characteristics of what might be termed "forest culture" and to use these characteristics as major categories for noting differences, although more research is needed to refine these and to relate them to other cultural factors (Hislop and Sims 1999). These zones reflect the climatic and ecological differences, the importance of forests in the national economy and as part of the landscape, and the way forests and trees relate to the culture through traditional uses, forms of recreation, folktales and legends.

The Northern Forest Culture covers Norway, most of Sweden, Finland and the Baltic states. Here the forest is a major element of the landscape, the national economy and the everyday lives of the people. Cities such as Stockholm, Oslo, Helsinki, Tallinn, Riga and Vilnius tend to be set within and surrounded by large tracts of forest which expand out into the greater surroundings. In the wider

Figure 5.3. Map of the different regions of European Forest Culture (after Hislop and Sims 1999).

countryside the forest is a major source of income and may be expanding naturally as agricultural use of land decreases. Valuable broadleaved trees such as cherry, ash and sycamore are not naturally occurring components of these forests except in the most southern parts, although they may be used in cities. If they do occur naturally, their presence may add a special dimension to the landscape.

The Central European Forest Culture covers those countries with a mainly continental climate but with milder winters and warmer summers than the northern countries. The landscapes are quite forested but not so heavily as in the north. The people of the cities use woods intensively for all kinds of recreation and the woods may be under considerable pressure. Forests have long been managed and traditional management practices, which originated in the forests of this zone, may be accepted as part of the landscape heritage. Forests and woods also occupy a key place in the rural landscape where cherry, ash and sycamore may be natural components of the countryside.

The Southern Forest Culture occurs principally in the Mediterranean countries as well as Portugal. The Mediterranean climate of hot summers and warm winters means that people live outdoors much of the time, therefore street trees, parks, squares and woodlands provide much needed shade. Forests may be significant elements of the landscape in mountainous or hilly areas where various forms of forest management, often of unique character to satisfy special markets, have been traditional, and where plantations of non-native trees have more recently been established. These countries have long developed urban cultures and the countryside is frequently intensively managed.

The North-western Forest Culture includes Britain, Ireland, Belgium, The Netherlands, Denmark, North-East France and Iceland. The common factor in these areas is the loss of most of the forest cover over the last three to four thousand years, so that forests now occupy a small percentage of the land area. Many forests are now of plantation origin and are intensively used. In current North-western popular culture, forests are sometimes regarded as alien places to which urban people are less spiritually connected than in the Northern zone, for example. These countries are also highly urbanised and industrialized with dense populations who need space for recreation. In Britain, Ireland and North-east France there are areas of traditional countryside that are a result of the initial clearings of the wildwood up to 5000 years ago (known as "ancient countryside" in Britain and "*bocage*" in France). In this countryside are densely hedged areas of irregularly shaped small fields containing small broadleaved woods, which are often managed by coppice systems and are highly valued as parts of the cultural landscape. Ash and cherry are frequently found in these landscapes, although sycamore is considered an introduced species in Britain and Ireland.

Public perceptions of forest landscapes

Foresters, farmers and other land managers tend to view woodlands differently from the general public. Farmers have historically cleared woodland in order to create farmland, especially in those parts of Europe where forestry is not an activity alongside agriculture, therefore there may exist a degree of antipathy from these people towards forests. The areas of woodland, which remained as the land was cleared, either fulfilled a useful purpose or were situated on inaccessible or useless terrain. Scientific forestry management started in the 17th century during the Age of Enlightenment and was only later applied in Britain, for example, in the 19th and, particularly, 20th centuries. Thus the foresters' taming of the remaining woodlands and the conversion or re-afforestation into tidy compartments of straight stemmed trees along rational, Cartesian lines has approximately 100 to 300 years of history, depending on the country. This is sufficient time for associations with such layouts, both positive and negative, to accumulate and become somewhat embedded in the cultures of different countries.

Against this background research, different from the perception studies noted above in relation to the Kaplans' work, has attempted to elucidate several aspects.

During the late 1980s and early 1990s research was conducted that aimed to find out what kind of forest or woodland landscapes were preferred by the public and to try to link this with design principles. Many of these studies were American but in recent years they have also been carried out in Europe in different circumstances; for example, general preferences for forest landscapes, both external scenic views and short distance internal views (Lee 2001), of specific forestry types (Breman 1995, Eleftheriadis and Tsalikidis 1990), of activities such as types of felling (Karjalainen 1996), forest stand structure (Koch and Jensen 1988) or re-afforestation of abandoned land (Hunziker 1995). Although much of this research focused on conifer landscapes of introduced species rather than native species, many of the general conclusions may still apply, including the preference for naturalistic patterns, diversity in the landscape as a whole and at the stand level, and landscapes with hills and water. The findings pertaining to stand structure are important for intensively managed forests of any species.

Research into the landscape value of farm woods attempts to calibrate design principles within woodland design guidelines (Bell 1998a, 1998b). This British study is the only one to attempt to do such calibration and is concentrated on broadleaved landscapes of four different types in the English lowlands. The conclusions are that people value patterns with no obvious geometry, prefer diversity and do not want the landscape to be overwhelmed by woodland when its current character is predominantly open (perhaps this latter conclusion is a reflection of the values in the North-west regional forest culture zone).

The benefit or value attached to making improvements in forests and woodlands has also been explored (Bell 1998a). This research concentrated on afforested conifer landscapes, therefore its application to broadleaved areas is questionable. It was found that removing geometric shapes, increasing species and structural diversity and moving towards continuous cover management created a positive value (calculated by contingent valuation methods).

These results demonstrate that in general the people of Britain and Ireland, and to some degree other parts of Europe, prefer forests and woodlands to look natural when seen externally as part of the wider landscape, and that they prefer more diverse, natural internal stand structure. These findings support the expert views of forest landscape architects and other forest landscape specialists as well as the guidance presented in publications on forest landscapes over the years around Europe.

Introduced species

Of the valuable broadleaved tree species under consideration, sycamore is considered an introduced species in Britain and Ireland. While there is enormous use of non-native species in many countries, they are frequently considered by certain sectors of the public to be more acceptable in parks, gardens and in city streets than in forests or the wider countryside. Sycamore is reviled by a number of conservationists because of its alleged tendency to colonize natural or semi-natural woodlands and cast heavy shade. While this is not strictly an aesthetic issue, it does affect people's preferences and the acceptability of using this species in certain locations.

5.1.3 The contribution of valuable broadleaved trees to the landscape

In many ways valuable broadleaved trees are no different from other species in the contribution they make to the landscape. Each species offers unique characteristics and the resulting pattern and composition of the assemblages of different stands, groups of trees or individual trees adds to the total aesthetic character to the landscape. This section will examine in more detail the particular contributions of ash, cherry and sycamore to the forest rural landscape and urban areas. Much of this section is unsupported by the research literature.

The forest landscape

Broadleaved landscapes, especially those of a range of site types and topography, are diversified by patterns created by tree form, texture, colour, height differences, crown spacing and seasonal changes. Often it is possible to recognise relationships in the forest pattern across a landscape by the underlying soil and relief. Mainly broadleaved landscapes occur over considerable areas of central, western and southern Europe. Elsewhere the dominant type may be conifers, where the broadleaves form a minor element, or mixed conifer and broadleaf forests, where the broadleaved stands may be significant components offering strongly contrasting textures and colours.

Ash tends to grow naturally on rich, moist soils, therefore it can be found in lower lying yet freely drained areas. The canopy is rather loose in form and texture which contrasts with oak, beech, sycamore and most other trees. The tree, having a thinner crown and coming into leaf later in the spring, allows light to penetrate to the ground so that the understorey of herbaceous plants can be rich, once more offering a contrast in the internal landscape to the shade created by beech or oak. The late flushing also means that the textural differences of the crown stand out longer as the other trees leaf out around them. Moreover, this creates interesting effects within the woodland as it is possible to move from shade to dappled shade and experience different temperatures and humidity on the skin. In the autumn ash continues to contrast with other trees because the leaves do not change to brown or gold, but suddenly fall from the tree. Thus, the forest canopy maintains colour diversity when the richer autumn colours are contrasted by the continued presence of subtle green.

Cherry is perhaps more often found in mixtures, small distinct groups and at the forest edge. Its main contribution to the landscape is in the spring when it is flowering, often together with blackthorn (*Prunus spinosa*) and hawthorn (*Crataegus monogyna*). This flowering is one of the most attractive elements of the forest at that time of the year before the leaves of most trees have come out. This frothy exuberance of flowers marks the true beginning of spring and the white presents a startling contrast with the fresh green of new grass or of cornfields nearby. Being a smaller tree than many of the others, cherry needs adequate space to flourish, hence its presence at the forest edge. Where there are open spaces within the forest, cherry can also be valuable at providing diversity along internal edges. This can provide enhanced aesthetic experiences during the spring. However, during the rest of the year it can be rather dull compared with its springtime glory, although the bark of mature cherry trees can take on a deep red-brown colour which is attractive at close range. In the autumn the leaves may briefly turn vivid red and add to the colour diversity of the season. The light branching habit permits some light to penetrate so that the ground surface is not as dark as beneath other trees. The presence of cherry at the forest edge also helps to grade the edge from the full height of taller oak trees down to shrubs.

In forests dominated by conifers, cherry may have an important role as part of the broadleaved component at edges in particular and during the early stages of regeneration after logging, when pioneer and light demanding broadleaves may constitute the first successional stage of the forest.

Sycamore has a distinctive rounded, fairly dense crown of dark, sometimes almost black, green in the middle of the summer. Being one of the first of the forest trees to flush in the spring it starts to emphasise the patterns, which may be much less pronounced later on, of the forest canopy at that time. By casting heavy shade to the forest floor, sycamore stands can be only sparsely vegetated beneath. The fallen leaves can create a substantial layer of litter in the autumn. At that time the leaves turn a dull bronze, not as bright as the leaves of Norway maple (*Acer platanoides*) but attractive nonetheless as part of the palette of forest colours.

From within the forest, stands of sycamore can be dense as growth competition is fierce yet shade tolerance is fairly high. This can result in stands of very straight trees, clean of branches and smooth, often a dull grey-green colour, which contrasts with the rougher bark of oak and the silvery grey bark of beech.

The rural landscape

In the more open conditions of the fields, hedgerows and small woods found in many rural areas, trees tend to take on very different forms and characteristics from those of the forest. Open grown trees have deeper crowns, fuller shape, are more highly branched and may be affected by browsing or damage from storms.

There are many different types of rural landscapes around Europe. In some areas, such as parts of Britain, Ireland and northern France, the enclosed hedgerow landscape is typical and provides opportunities for all three tree species to grow with full crowns. Some of these landscapes contain so many hedgerow trees that they appear almost to be wooded. Other areas, for example the Northern European Plain, are generally more open and the enclosure in the landscape is provided more by woodlands, together with clumps of trees next to farmsteads or along roadsides and watercourses. The most intensively farmed areas may have few trees and present large scale open vistas. Increasingly, these intensively managed areas are becoming dominated by man-made structures such as masts, large buildings, power lines and transport infrastructure. Planting trees at a suitable scale can help to reduce the impact of these structures. Where the soil is rich, ash, cherry and sycamore, together with poplars (*Populus* spp.), may have a significant role to play because they grow quite quickly and last a long time.

The rural landscape, especially hedgerows or field edges, may present opportunities to grow valuable broadleaved trees because they can be planted with wide spacing to provide ample room for crown development. The practice of high pruning may result in trees of different proportions to those normally found in such landscapes. Yet, as long as they are interspersed with trees of different growth character, this aesthetic concern can be accommodated. Farmers could potentially find growing trees a valuable means of diversifying their production.

Ash can be found as part of the rural landscape, possibly because of its popularity as firewood or because it casts less shade than other trees. The characteristically loose, heavily branched form achieves greater development. The late flushing and relatively light canopy enable tantalising filtered views to be gained through the tree while other species block views entirely. Ash is often associated with particular landscape types. It is a common tree, for example, in magnesian limestone areas of Derbyshire, England. Its grey bark and light green leaves blend closely with the light grey stone used for field walls and buildings, thus maintaining a unity in the landscape. Ash is reasonably fast growing and contributes to the landscape quite soon after it is planted.

Cherry, as noted above, gives its greatest contribution to the landscape in the spring. It can be found along many hedgerows and at the edges of small woods

Figure 5.4. In this rural landscape in Wales there are many trees in hedges and woods. Ash, sycamore and cherry are native or grow abundantly here and could be cultivated by farmers as a valuable crop. Photo: Simon Bell.

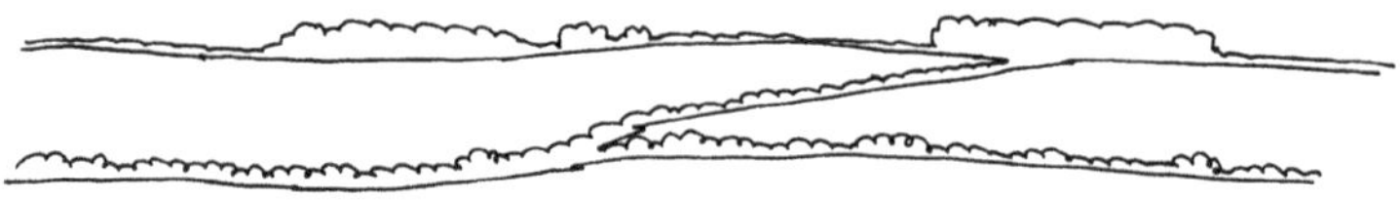

Figure 5.5a. A hedgerow landscape where the planting of valuable broadleaved trees is being considered.

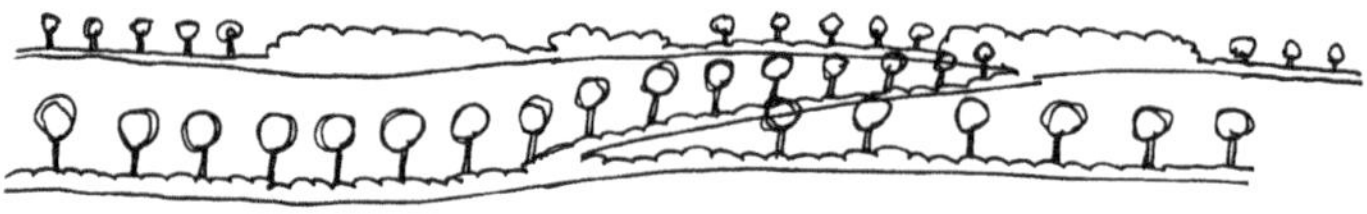

5.5b. Trees of the same age, regularly spaced look artificial and have a negative visual impact.

5.5c. Spacing the trees irregularly looks more informal and natural in the landscape.

5.5d. Mixing ages of trees provides a continuous means of production and also looks better in the landscape.

where, if the landscape is small in scale, it has a major effect. Cherry, blackthorn and hawthorn in flower together make a truly unforgettable sight in the countryside. Cherry tends to have quite a round but not very dense crown and it being a small tree provides structural contrast along hedges or woodland edges, which also helps keep the scale more intimate. The russet colour of the autumn leaves adds diversity, depending on its importance in the landscape. Cherry is commonly found in fields, along roadsides, in country gardens and on village greens. Its smaller stature ensures its place in the village scene.

Sycamore grown in the open develops a full, dense, round canopy, which casts heavy shade. This can be a drawback in farmland as it may affect the crops grown beneath it. This crown density, coupled with its tolerance of more severe climatic conditions than many other broadleaved trees, has made sycamore a feature of coastal and upland areas where it can provide some shelter. Many farmsteads in exposed upland areas of Britain are characterised by groups of or single sycamore trees positioned to provide shelter from storms. This can soften and humanise the otherwise often bleak and inhospitable setting of such houses. Sometimes the trees are severely sculpted by the wind, which may emphasise the sense of bleakness.

The urban landscape

Trees in the urban landscape may be found in woodlands, parks, along streets and in gardens. Their role is usually rather different from the forest or rural landscape. In particular, their aesthetic functions and properties are usually of great if not the most importance (Bell et al. 2005). These aesthetic functions are not confined to the decorative properties of trees, but include the architectural use of trees arranged in different ways to create space, mass, and enclosure, to control vistas and soften the hard forms of buildings. Other functions may include providing shade, shelter, dust filtration and noise abatement.

Urban woodlands are valuable resources for recreation, therefore their layout and management is often primarily concerned with providing a rich, accessible, safe environment. Different approaches to management are likely depending on the type of forest culture associated with different cities. In the North-West European Forest Culture strongly enclosed forests are not preferred, therefore open spaces and open stands are often desirable features. In the Central European Forest Culture denser, more extensive woodlands are more accepted.

Ash, cherry and sycamore may play the same roles in urban woodlands as in the larger forests and rural landscapes. The growing environment may be similar, although growth in the soil and climatic conditions may be difficult in places. Since sycamore is hardy, easy to establish and resistant to exposure and salt, it may be a popular choice in urban woodlands. However, its heavy shade may require it to be planted in mixtures of less densely crowned tress such as ash. Cherry provides the spring attraction, which helps to mark the seasons in an environment where they are less pronounced than in the countryside.

Park landscapes generally use trees in small wooded areas, clumps, groups and as single trees, in layouts both highly formal and deliberately informal. A wider range of species may be employed, especially for specimen trees, such as those

Figure 5.6. A city park in the Hague, the Netherlands. There is a role for ash, sycamore and cherry in such places and it might be possible to grow then for good quality timber as well as to fulfil their role in the design. Photo: Simon Bell.

introduced for their form, foliage, flowers or autumn colours. Sycamore, ash and cherry also have a role to play. Ash and sycamore may be used as matrix elements of larger blocks or clumps to give enclosure and to let light into the stands. Cherry may be used extensively, although the native species (*Prunus avium* and *Prunus padus*) may be heavily outnumbered by Japanese and other more colourful species or varieties. In some areas there is a movement to use native species wherever possible and this may lead to an increase in the use of native cherries.

Street trees are used not only to provide shade and shelter, but also to create deliberate architectural effects such as avenues or geometric arrangements to reflect the layout of city streets. These trees must fulfill certain management and safety requirements. Street trees hardly ever include ash as its form and growth habits are not suitable along streets and it is not amenable to pruning, compared with lime (*Tilia spp.*) for example. Cherries, usually ornamental varieties, are popular in such places. Many of these are grown from grafts and only grow to limited dimensions, which is an advantage in such areas. Native species are generally undervalued and could be used more often. They can often be formed by pruning and can provide a variety of size of tree between the restricted ornamentals and the large forest trees. Sycamore may be used along some streets, but its tendency to produce lots of seed can be a problem. Its hardiness, on the other hand, is a virtue in the less optimal conditions of many urban soils and microclimates. However, sycamore may be unpopular along residential streets because of its shade and seeding habits. It nevertheless provides an extra option to the usual choices such as lime, plane (*Platanus spp.*) or horse chestnut (*Aeschylus hippocastanum*).

It should be technically possible to grow tall, straight valuable broadleaved trees in urban conditions where space is available. In some locations, the high canopy and dramatic pattern of straight stems can produce a significant architectural effect.

Figure 5.7. In a city street it is advisable to grow trees with high crowns and at wide spacing. This means that with could management they could also produce quality timber.

Urban trees need to be managed for safety and usually do not have a long life span, therefore removal of such trees may not provoke a huge public reaction. Trees may be grown in wide streets, in parks and in urban woodlands. Intensive management and pruning are part of the management regime of urban trees which is conducted by staff who have the appropriate expertise and equipment.

5.1.4 The implications of the intensive silvicultural management for the landscape

When any species of tree is grown under intensive management conditions in order to increase both timber volume and stem quality, there are implications for the landscape. Silvicultural methods may produce visual effects that people find intrusive. These may include the resulting patterns of tree stands in the wider landscape or the internal arrangement of planted areas. In order to enable forest managers to continue to produce timber efficiently while protecting the landscape, a number of visual design principles have been developed and applied over the last three decades. These principles, based on landscape architectural practices, were first established in the UK, but since the early 1990s have been applied in many other countries in Europe and elsewhere (Bell et al. 2005, Bell 2004, Bell and Nikodemus 2000, Forestry Commission 1992, Lucas 1991). Since many of the potential visual problems of intensive silviculture can be avoided by applying these principles, as opposed to following rules or regulations, some of the main principles are briefly described here. Some of the research referred to earlier has validated the application of these principles (Bell 1998b).

Shape

One of the main ways that we see patterns is through recognition of shapes, and the interactions between different shapes. We can often recognise differences between natural and artificial shapes, and between ones that fit into a particular scene and those that do not. Natural shapes tend to be irregular, asymmetric, have diffuse edges and variable size. Artificial shapes are often geometric, symmetrical and hard edged. Shapes that have a major influence in the landscape are those of landform and vegetation. Landform shapes can be angular or rounded, simple or complex. Natural vegetation patterns frequently reflect those of landform and can also arise from natural disturbances such as forest fire or wind throw. Such vegetation shows a marked contrast to the artificial shapes formed by square felling coupes, straight roads and plantation patterns.

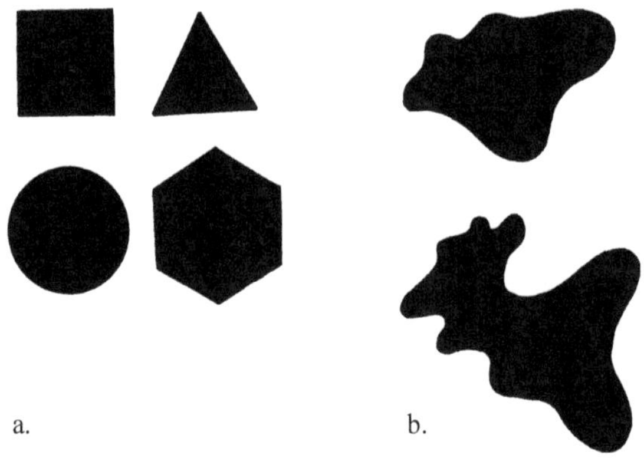

Figure 5.8a. These are examples of typical geometric shapes, which we associate with the man-made.
5.8b. These shapes are organic and appear more natural. (Source Bell 2004)

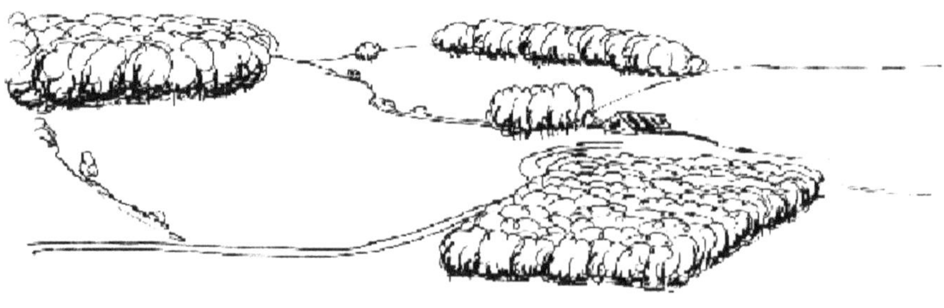

Figure 5.9. In this sketch the rectangular woodland block looks out of place in the scene where everything else is more organic in shape (Source Bell and Nikodemus 2000)

Visual force

In addition to designing shapes that reflect natural ones, in cases where landform is prominent they should reflect the underlying structure because of the way in which landforms direct our gaze. It can be demonstrated that we tend to see ridges as having a downward, and valleys an upward, direction or flow. The pattern of ridges and valleys can be plotted on a contour map and overlaid on a photo or sketch of the scene. Shapes that appear to contradict these flows or forces look out of place while those that reflect them appear more harmonious. It is on horizons that these forces can be most prominent, therefore forest boundaries that cross the horizon require extra care in their design. Landform can be used in areas of subtle topography; it is only in completely flat areas that it is irrelevant.

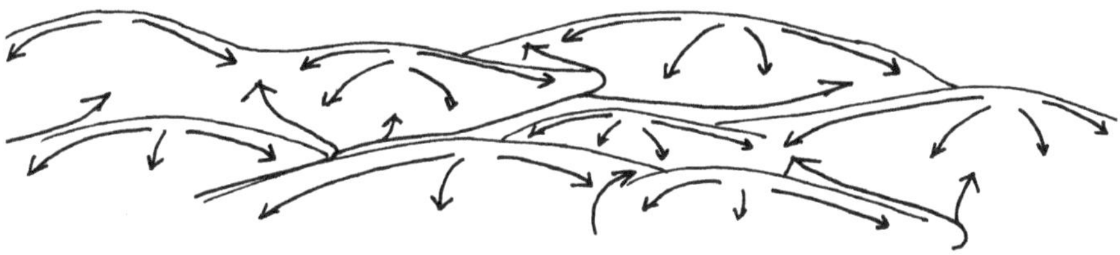

Figure 5.10a. This simple sketch illustrates the way that visual forces are perceived to flow down ridges and up valleys.

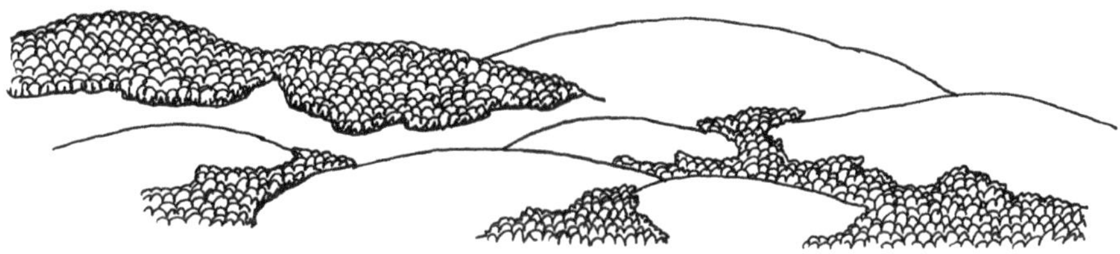

5.10b. The same landform with woodland planted to respond to the visual forces.

Scale

Scale takes into account the size of elements in a landscape relative to the size of the human figure. The scale of the landscape depends on the relative magnitude of the landform and on the distances over which we can see. As a general rule, patterns should be of large and small scale in respective large- and small-scale landforms. Scale also tends to increase from valley bottoms and lower slopes, to hill tops and mountain peaks.

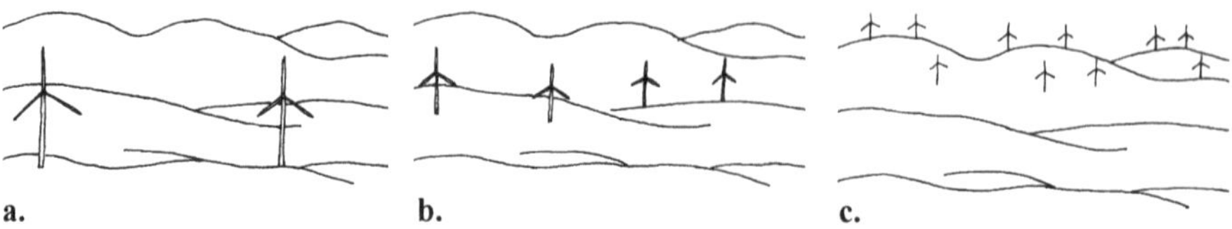

Figure 5.11. Landscape scale depends on the presence of the human figure or objects of a known dimension. a. appears as a small scale landscape because the hills are a similar scale to the wind turbines, while in b. the scene is medium scale and in c. large scale, because the turbines are much smaller in comparison to the hills.

Enclosure

Enclosure occurs when views are contained and space is created by elements such as forest or trees. This effect does not require complete physical enclosure, as even partly enclosed spaces appear separate from those beyond. The degree of enclosure is a combination of the height of the enclosing element and the distance over which it is perceived. Thus, tall trees at some distance from the observer can produce the same sensation as smaller closer trees. Overlapping forest edges may produce apparent enclosure and reduce the appearance of open space, while in reality these pieces of forest may only occupy a small physical area compared to the size of the space enclosed. This is a very important illusion in flat and undulating landforms (Figure 5.12). The use of enclosure can also be used to prevent the entire landscape from being visible at once, thus maintaining a sense of mystery.

Figure 5.12. The organic black shapes enclose a space within them, though it is not completely cut off from outside (Source Bell 2004).

Figure 5.13. In this flat landscape the trees and woods partly enclose the space and have the effect of reducing the scale of the landscape (Source Bell and Nikodemus 2000).

Nearness

When elements such as trees, clumps or small patches of forest are scattered widely apart we tend not to see any spatial relationship between them. However, if they are positioned quite close to one another, we tend to see them as belonging to a group. This spatial characteristic is important for the creation of apparent masses of trees or for creating cohesion when using small groups or areas of trees in a larger landscape.

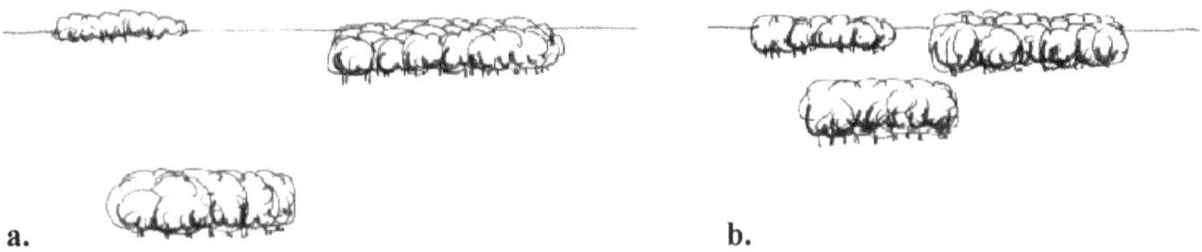

a. b.

Figure 5.14a. The three small woods are positioned too far apart to appear to belong to the same group.
5.14b. The same three woods are near enough together to read as a group and thus they also seem to have a larger scale (Source Bell and Nikodemus 2000).

Interlock

Shapes that interlock with each other tend to produce a more coherent pattern than those that meet abruptly. The individual shapes appear as parts of a single composition. It is easier to achieve interlock with irregular organic shapes and, where landform is pronounced, those related to topography through the application of visual force.

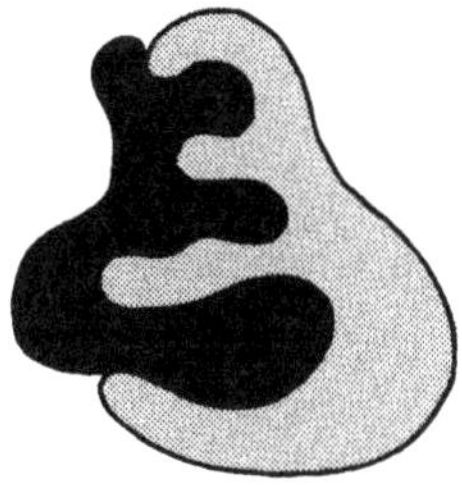

Figure 5.15. These two shapes, black and grey, interlock with on another to form a single element (Source Bell 2004)

Figure 5.16a. This open area enclosed by forest is bland and uninteresting in part because it is not interlocked with the forest.

5.16b. The space is interlocked with the forest and so appears to be an integral part of it (Source Bell and Nikodemus 2000).

Coalescence

Elements such as trees or patches of forest that are physically separated by wide distances may, nevertheless, be visually connected in a landscape to give it a greater forested appearance. This is because all the features coalesce together to produce a single visual mass. This principle can be used to produce greater coherence among many smaller, isolated, small-scale elements, such as clumps of trees, or to lose the sense of openness otherwise present in a landscape.

Figure 5.17. These three small woodlands are positioned in such a way that they coalesce in the view and appear to become on element, so increasing their apparent scale (Source Bell and Nikodemus 2000).

Proportion

In order to maintain a generally open appearance in the landscape it is necessary to maintain the visual proportion of forest below one third of the scene. The sense of openness is lost when two thirds or more of the visual area is wooded. This is a complicated yet important concept to understand in order for landscape design to achieve the aims of controlling the scale and rate of landscape change. The application of enclosure, nearness and coalescence can be used to create the illusion of more or less forest than in actuality and thus affect the apparent proportion.

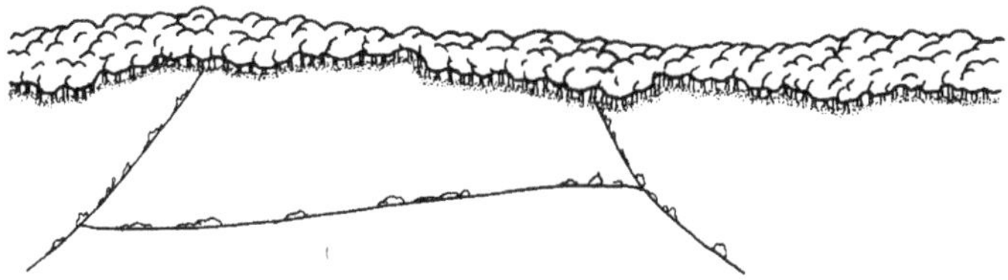

Figure 5.18a. In this scene the woodland occupies less than ⅓ of the visible proportion and looks out of balance.

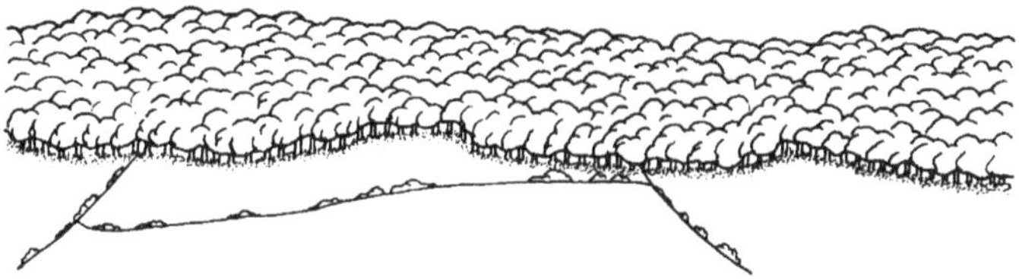

5.18b. Here the woodland occupies around ½ of the scene and also looks unbalanced.

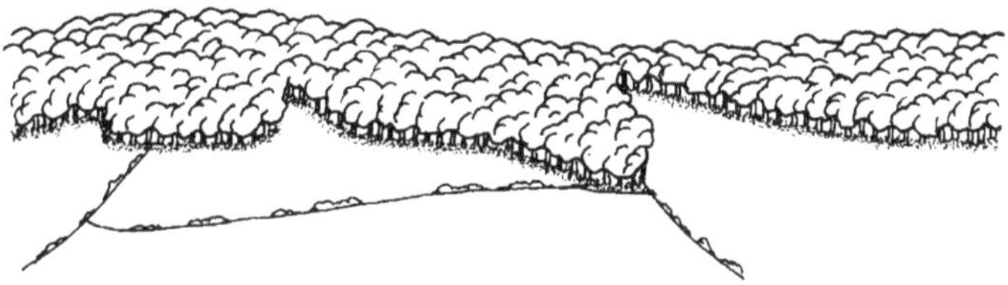

5.18c. In this version the woodland appears to occupy around ⅓ of the visible scene and appears balanced (Source Bell 2004).

Unity

In order to achieve coherence, all the constituent elements of the landscape must relate to one another and appear to be part of the scene without any jarring features. Similarity of forest shapes, especially where they relate to landform, is a particularly important feature. The use of enclosure and interlock are appropriate in flatter landscapes. Similar colours, textures and edges are also beneficial in creating this effect, although the presence of some eye-catching and contrasting features can also be important in bringing a scene to life. Such features may be man-made, such as buildings or structures, while occasional individual trees of character may provide a focal point in a scene (such as ancient trees found near old farms). Too much variety can cause the scene to become visually chaotic.

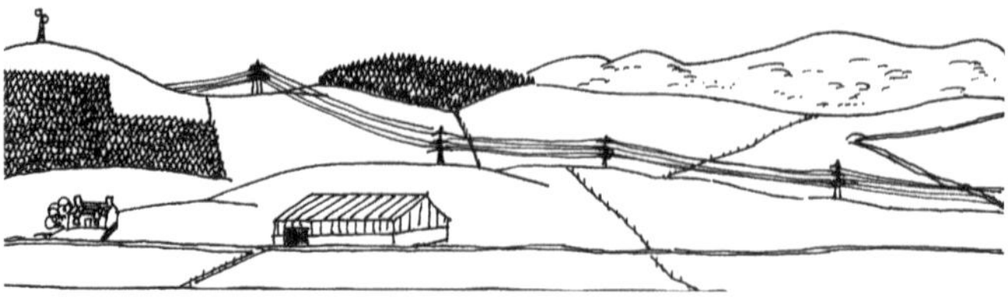

Figure 5.19a. This scene is disunified because there is no harmony between the different elements and their shapes, position, relation to landform, scale or proportion.

5.19b. Appears much more unified because there are no jarring elements and those present relate to each other and the landform.

Diversity

As noted above, some contrasting elements help enliven a scene as we desire a sense of complexity as well as coherence. Thus simple scenes with insufficient visual challenges can be bland and boring. Within a threshold, the more variety there is, the more interesting the landscape is likely to be. However, the elements of diversity should be appropriate to the landscape and not induce a sense of visual competition amongst them. It is important not to reduce the expression of diversity across a country or over the whole of Europe by introducing too much standardization in forest specifications. The regional and local expression of diversity should be retained through the careful analysis of landscape character so that each design is developed to reflect and enhance a landscape of strong and appropriate character. At the local and site level there may be a deeper sense of identity or spirit of place reflected (Figure 5.20).

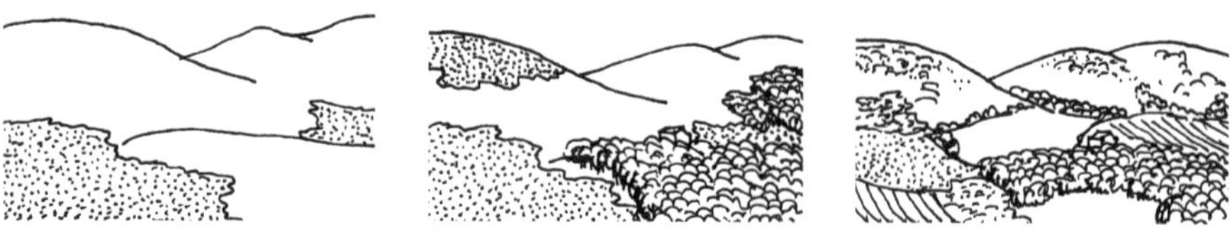

a. **b.** **c.**

Figure 5.20a. The landscape is not very diverse and as a result is not so interesting to look at. **5.20b.** The scene is more diverse, containing more elements (while also remaining unified), while **c.** is even more diverse, though not too diverse – visual chaos has not yet occurred and the landscape remains unified – and so is even more interesting to look at. However, adding more elements to this scene might tip the balance so it becomes too diverse and also begins to lose unity.

Spirit of place

Many places are identifiable due to a presence of key features that give them a unique sense of place. These may be natural (rocks, water, unusual landforms, ancient trees) or cultural (archaeological features, old churches, castles, unique crop systems, associations with artists or historic events). It is important that the particular spirit of place is identified so that any expansion or management of forest or woodland does not detract from or compromise it. Well-designed forests can be used to frame and emphasise some landscape features and increase their sense of mystery. Unsightly features can be screened out and the eye may become focused on key elements at certain viewpoints.

The next section applies these principles to a number of aspects of forest layout in relation to intensively grown valuable broadleaved trees (Forestry Commission 1992). This section is the most speculative and is based largely on the expert application of the design principles to the subject, grounded in the experience of practical work. There is scope for interesting research to test whether the suggestions for design presented in this section work in practice and whether there are any cultural differences that affect them.

Plantation layout

Where a plantation of valuable broadleaved trees is to be created in an open agricultural landscape, one of the first concerns should be of the shape and scale of the area. The shape should reflect the dominant characteristics in the landscape, such as landform or field pattern. If the landscape is enclosed by hedges and hedgerow trees and contains small woodlands, it is easy to select suitable fields for planting. If the fields are irregular in shape, a strong sense of unity will be achieved if the plantation boundary follows those of the fields. Where the landscape is mainly open and landform is a significant factor, the designer should achieve a balance between shapes that responds to the landform and reflects the typical existing woodland elements. However, if woodland cover is low and the existing pattern weak, it is better to follow landform. Plantation shapes should not be too complicated, especially if the newly planted trees have to be protected by fencing. However, a geometric fence layout can be used regardless of the shape of the planted area within.

The size of the planted areas should reflect the scale of the landscape, so that larger plantations will be appropriate if the landscape is broad and sweeping. If the individual plantation units are necessarily small, the use of nearness, enclosure and coalescence should help to achieve unity. The balance between forest and open landscape should also be considered, especially in relation to the various points from which the scene may be viewed, such as from a road. Planted areas closer to a road may block views and create the illusion of a larger planted area than may be the case.

The shape of plantations set in a matrix of existing forest should follow the same principles, especially if the forest has a somewhat natural stand structure of irregular shaped areas comprised of different species closely related to soil type.

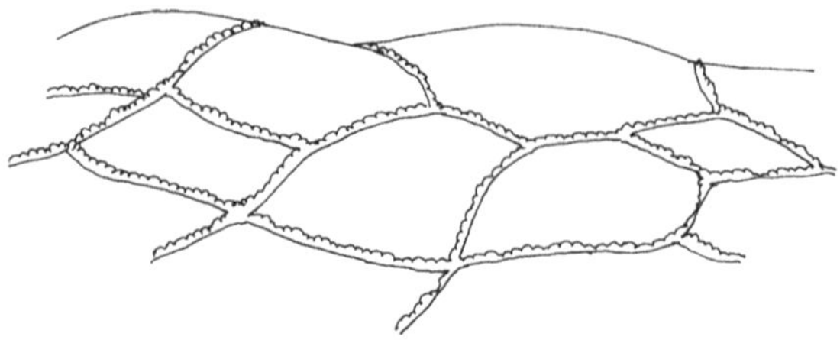

Figure 5.21a. A landscape of moderate topography and some enclosed hedgerow-bounded fields.

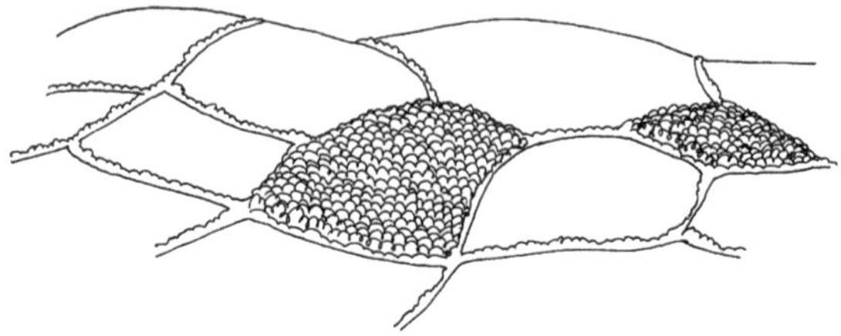

5.21b. Some fields are planted with woodland and remain unified by the presence of the hedgerow pattern, although the proportion is a little under ⅓ of the scene.

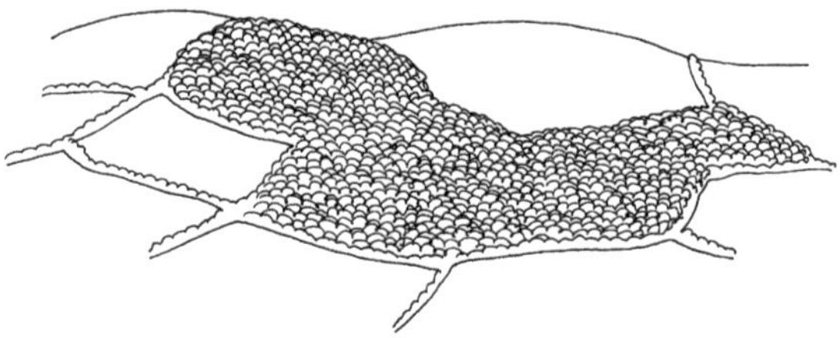

5.21c. More fields have been planted and the character of the landscape has started to change. More woodland may alter the balance between wooded and open and would need careful thought.

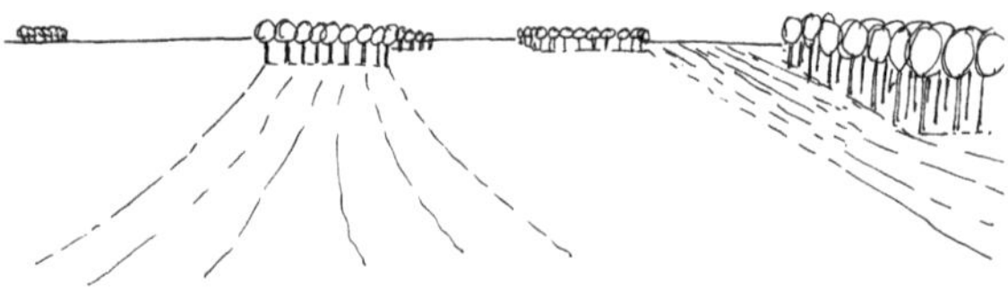

Figure 5.22a. In this open landscape some small stands of valuable broadleaved trees have been planted but their position makes them appear to float because they are not near enough to each other, nor do they coalesce.

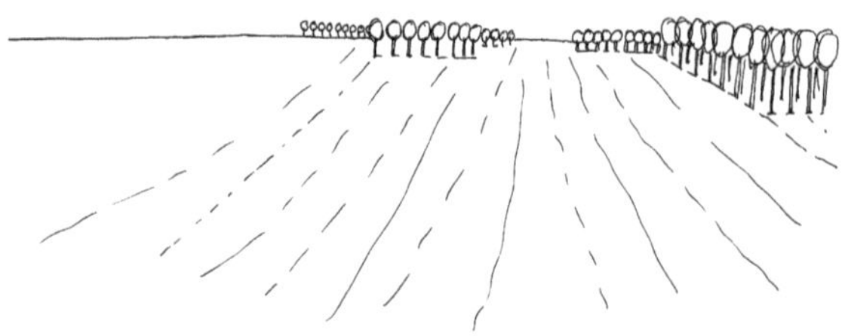

5.22b. The stands of trees are positioned so that they appear closer to one another and they are also coalescing and thus achieve a better scale and proportion.

Rectangular shapes, especially if visible on a hill or mountainside look disunified and may contradict landform. Shapes should therefore reflect the character of the landform and stand types so that they remain unified. The scale and proportion of plantations should also reflect that of the landscape. Plantations should ideally form no more than two-thirds of the visible proportion of the forest. The same issues apply to the compartments within a plantation. These traditionally have been rectangular in shape, although this can be as intrusive as any other geometric shape in a landscape of organic patterns. Linear open spaces used to delineate compartments only exacerbate the problem.

The recognition of a plantation seen from within the forest, likely means that the overall shape has not registered. Instead, the details of the layout are likely to create a greater impression. Simple row planting of intensely managed trees of the same age may present a strong sense of coherence but lack complexity and be largely devoid of mystery. Linear, straight rows of trees are also highly unnatural and represent human control over nature. If possible, given the terrain and silvicultural techniques employed, linear rows should be avoided and the spacing of trees should be irregular. If plants of clonal origin are used, this can also present an artificial effect, therefore random mixtures from several different clone origins should be used. Geometric shapes in the layout will be obvious if linear compartment boundaries, which may also function as recreational routes, for example, have been used. In such cases the boundaries should not only be organic in shape, but their width and edge structure should also vary.

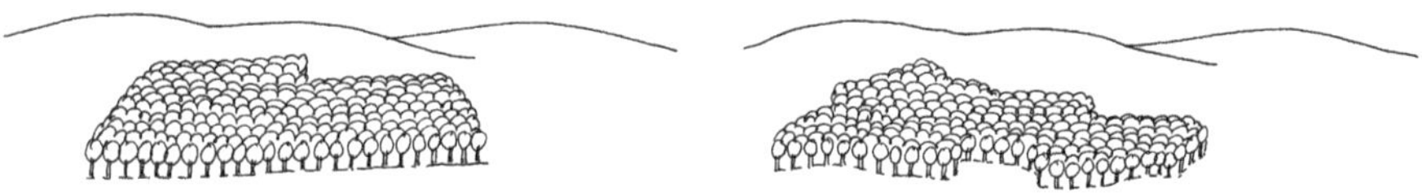

a. b.

Figure 5.23a. This plantation in this open landscape looks out of place because of its geometric form and lack of relationship to the landform.
5.23b. looks much more comfortable – it is organic in shape, with flowing lines that reflect the landform shape, as well as responding to the visual forces in the landscape.

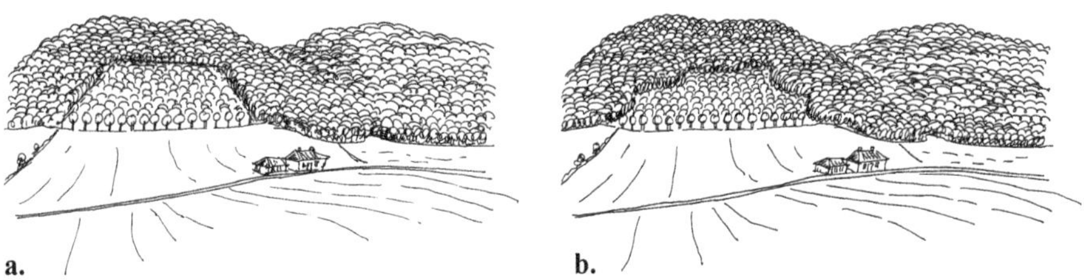

a. b.

Figure 5.24a. The plantation of valuable broadleaved trees in this forest area looks out of place because of its geometric shape and lack of relationship to the landform.
5.24b. looks better because the shape borrows from the surrounding landscape and reflects the landform.

Figure 5.25. This plantation of cherry shows the formality of a regular grid pattern.

a. d. b. e. c. f.

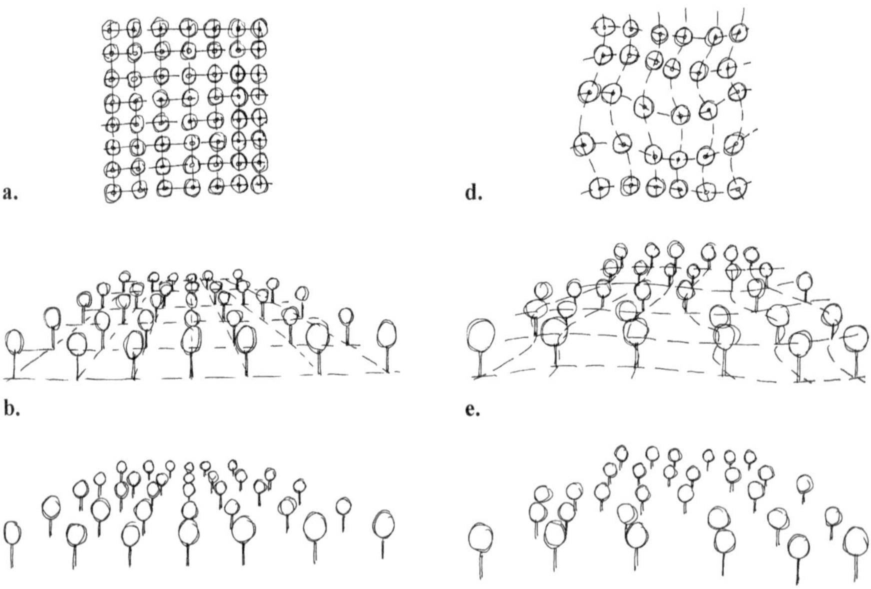

Figure 5.26. A plan view of a grid layout typical of plantations and orchards (a). Seen in perspective this grid produces a strongly geometric effect (b) which remains evident when the grid lines are not visible as in c. However, when a distorted grid is used, shown in plan (d) the appearance in perspective is much less artificial (e and f) yet the arrangement remains practical to manage.

Stand structure in plantations is usually much simpler than that found in natural forests. The optimum spacing for timber production may translate into sparse shrub and herb layers. However, with trees like ash, which cast less shade than oak, beech or most conifers, a richer understorey and ground layer should be possible. This will give at least some chance for spring flowering plants to flourish beneath even the most intensively managed stands. Sycamore casts more shade and allows only a very impoverished ground layer to persist. However, if the stands are regularly thinned as they age, it should be possible to allow more of an understorey to develop, or even advanced regeneration to begin. This will increase the aesthetic attractiveness of most forest stands, adding to their complexity and mystery.

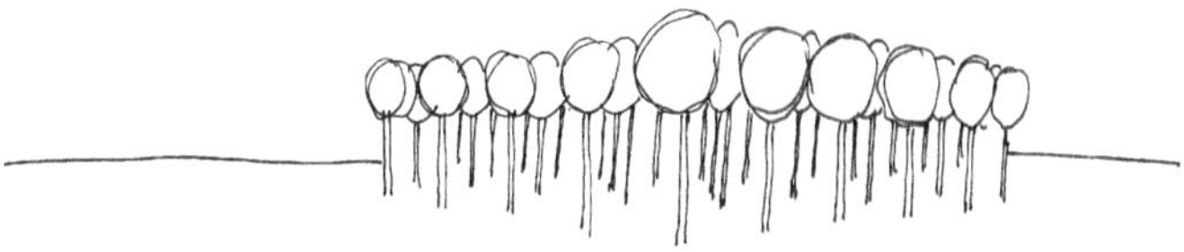

Figure 5.27a. The simple structure and wide spacing of a plantation of valuable broadleaved trees lacks interest and does not resemble a natural forest.

5.27b. The addition of understorey and edges of shrubs and other plants improves the visual effect.

While thinning, especially selective, improves the internal appearance of the plantation, final cuttings have an effect that many people find unattractive since clear cutting causes such a dramatic change to the landscape. Moreover, if plantation-grown trees are cut at an earlier age than normal stands, the rate of change of the landscape may be greater. Although the initial shapes, even those originally geometric, of the planted areas may soften over time, they once again become highly visible after felling. Thus, if older previously established plantations are ready for felling, it may be desirable to redesign the felling shape to avoid creating a geometric pattern that will persist into the next rotation. Hard edges can also be reduced by selectively thinning along the coupe margin into the adjacent stands. If shelterwood systems are used, the coupe shape may not be as evident during the early phases, such as during seeding fellings, but will become more evident during later phases of overstorey removal. Other continuous cover silvicultural systems may not be suitable for valuable broadleaved trees such as ash, cherry and sycamore, but if used generally have a lower impact.

Natural forests generally consist of a number of species in a mixture. This is not usually an intimate mix of all species, but rather a fine scale mosaic of different sized groups of each species located according to the micro-site characteristics. This contrasts with mono-cultural plantations, unless such mixtures are purposely planted. While silviculturally more complicated, such stands are more natural and attractive from both interior and exterior viewpoints.

5.1.5 Conclusions

Valuable broadleaved trees such as ash, cherry and sycamore are not found universally over Europe. Their contribution to the landscape varies from place to place, such as in the forest, wider countryside and urban environment. Forest landscape aesthetics as a well-developed discipline can provide practical guidance to forest managers who may wish to engage in intensive silviculture in order to maximise timber quality and volume while avoiding intrusive effects. This guidance is in the form of a series of design principles validated to some degree by research and tested by several decades of practical application in the field of forest management, especially in countries with a history of plantation forestry such as Britain. However, these principles should be interpreted for each specific landscape wherever trees are being planted. Such application ensures that the inherent landscape diversity found across Europe is respected and that the different regional forest cultures are reflected. Since the approach suggested in this chapter is yet to be tested on the ground, there is scope for research into the potential visual impacts of different layouts. This research could use visualisation to test alternatives on a large sample of the populations of different European countries where valuable broadleaved trees are likely to be grown.

References

Bell, S. 1998a Woodland in the landscape. In: The landscape: past and future perspectives. Atherden, M.A. and Butlin, R.A (eds.) The Place Research Centre, University College of Ripon and York St John, York.

Bell, S. 1998b Landscape values of farm woodlands. Information note 13, Forestry Commission, Edinburgh.

Bell, S. 1999. Plantation management for landscapes in Britain. International Forestry review 1(3).

Bell, S. 2000 Landscape: pattern, perception and process. Sponpress, London.

Bell, S. 2004 Elements of visual design in the landscape. Second Edition, Sponpress, London.

Bell, S. and Nikodemus, O. 2000 Handbook of forest landscape planning and design. State Forest Service, Riga, Latvia (in Latvian).

Bell, S., Simson, A., Blom, D. and Rautamäki, M. 2005 Design of Urban Forests. In: Urban Forests and Trees.A Reference Book. Konijnendijk, C.C.; Nilsson, K.; Randrup, T.B.; Schipperijn, J. (eds.). Springer. 516 p.

Berleant, A. 1982 The aesthetics of environment. Temple University Press, Philadelphia.

Breman, P. 1995. L'analyse visuelle du paysage forestier et les conséquences possibles sur l'aménagement et la gestion. Bulletin Technique de l'Office National des Forêts 28: 31–38.

Brush, R.O. and Shafer, E.L. 1975 Application of a Landscape-Preference Model to Land Management. In: Landscape Assessment: Values, Perceptions and Resources. Zube, E.H. Brush, R.O. and Fabos, J.G. (eds.) Halstead Press, Pennsylvania.

Bryson, B. 1998 A walk in the woods. Broadway Books, New York.

Burke, E. 1958 A philosophical enquiry into the origins of the sublime and the beautiful. Boulton, J.T. (ed.) Routledge and Kegan Paul, London.

Eleftheriadis and Tsalikidis 1990. Coastal Pine Forest Landscapes: Modelling Scenic Beauty For Forest Management. Journal of Environmental Management 30.

Forestry Commission 1989 Forest landscape design guidelines. Forestry Commission, Edinburgh.

Forestry Commission 1992 Lowland landscape design guidelines. HMSO, London.

Haber, W. 2002 Kulturlandschaft zwischen Bild und Wirklichkeit. Schweizerische Akademie der Geistes- und Sozialwissenschaften. Bern.

Hislop, M. and Sims, S. 1999 An examination of forest recreation policy in four European forest cultures, focussing on the impact of silvicultural practice. Unpublished paper. Forestry Commission Research Agency, Edinburgh.

Hoskins, W.G. 1995 The making of the English landscape. Hodder and Stoughton, London.

Hunziker, M. 1995. The spontaneous reforestation in abandoned agricultural lands: perception and aesthetical assessment by locals and tourists. Landscape and Urban Planning 31: 399–410.

Kaplan, S., Kaplan R. and Ryan, R. L. 1998 With people in mind: Design and management of everyday nature. Island Press, San Francisco.

Karjalainen, E. 1996. Scenic preferences concerning clear-fell areas in Finland. Landscape Research 21(2): 159–173.

Koch, N.E. and Jensen, F.S. 1988 Forest recreation in Denmark Part IV The preferences of the population in Det forstlige Forsøgsvaesen I Danmark Vol XLI., Trykt I Kandrups Bogtrykkeri, Copenhagen.

Lee, T.R. 2001 Perceptions, attitudes and preferences in forests and woodlands. Forestry Commission Technical Paper 18, Forestry Commission, Edinburgh.

Lucas, O.W.R. 1991 The design of forest landscapes. Oxford University Press, Oxford.

Rackham, O. 1986 The history of the countryside. H.M. Dent and Sons, London.

5.2 Effects of Management of Valuable Broadleaved Trees on Nature Conservation

Albert Reif
Institute of Silviculture, Site Classification and Vegetation Science, University of Freiburg, Germany

Abstract

Forestry with valuable broadleaved tree species is highly productive and protects soil and ground water. Evaluations of forests with valuable broadleaved tree species depend on the previous vegetation, the site and subsequent stand treatments. Conflicts may arise with nature protection, mainly on the genetic and landscape levels:

- Forestry induces shifts of genetic composition and losses of genetic diversity of forest trees, including valuable broadleaved tree species. Due to the genetic selection of trees in production forests, a sufficient number of large forest reserves must be created in order to preserve the site-adapted genetic diversity of the tree species.
- Mixed stands can exhibit greater tree species diversity than beech dominated stands. The present situation of even aged uniformly sized valuable broadleaved forests can be improved through silvicultural measures thereby resulting in uneven mixed forest structures.
- Harvesting of the individual trees after ca. 70 to 100 years hinders the development of large-timbered stems and impacts fauna that depend on old trees for survival. Only forest reserves have the potential to provide terminal and decaying phases of forests with temporal and structural niches for those species which are absent in production forests, e.g. species inhabiting large diameter decaying wood.
- Forests with valuable broadleaved tree species may result form melioration (drainage) of extreme sites and subsequent altering of their original tree composition. From the viewpoint of nature conservation, extreme sites and their biocoenoses have to be preserved, even if they are unproductive.
- Many afforestations are performed with valuable broadleaved tree species. Afforestation of sites with valuable biocoenoses must be avoided, e.g. oligotrophic grassland. Compensation measures for environmental destruction, outlined in environmental impact assessments, should increasingly take natural succession into consideration as opposed to afforestation with genetically dubious material.

Keywords: valuable broadleaved tree species, naturalness, originality, rarity, endangerment, diversity

5.2.1 Introduction

The category "valuable broadleaved trees" includes a number of deciduous species, which have particularly valuable timber, e.g., ash (*Fraxinus excelsior*), maple (*Acer pseudoplatanus, A. platanoides)*, lime (*Tilia platyphyllos, T. cordata*), elm (*Ulmus glabra, U. laevis, U. minor*), cherry (*Prunus avium*), the *Sorbus*-species *S. aria, S. torminalis* and *S. domestica*, and the alder species *Alnus glutinosa* and *A. incana*. On sites with high water and nutrient supplies these trees are able to grow over and even out compete the shade tolerant "climax" species beech (*Fagus sylvatica*) and fir (*Abies alba*). Due to an early culmination of growth, valuable broadleaved tree species may gain dominance over the "climax" species during the initial decades following stand establishment (Ellenberg 1996). In recent years, the cultivation of valuable broadleaved trees has increased in many European countries. The underlying changes in economy and land use affect the sites, the successional processes, the situation of associated species, and the landscape character. The cultivation of valuable broadleaved trees must therefore be analysed and evaluated not only in terms of economics, but also ecology and nature conservation.

Valuable broadleaved tree species tend to be planted on relatively moist, fertile soil. On such sites they form a near-natural canopy composition because they also grow there naturally. Conflicts can arise when afforestations alter cultural landscapes and grow over rare man-made biocoenoses, when hydromeliorations endanger rare species and habitats, and when owners of productive forests with valuable broadleaved tree species on drained, previously wetland sites oppose restoration of wetland forests. Examples include the debates about how to construct and use the flood retention basins and their forests along the Rhine River (Volk 2001, Reif 1997).

5.2.2 Which criteria of evaluation in forest nature preservation should be given high weight?

The evaluation of valuable broadleaved trees in production forests should be based upon suitable criteria and the relationships between them. In forested landscapes the criteria "preservation of soils", "naturalness and originality", "rarity and endangerment" and "diversity" should be more heavily weighted than other criteria (Reif et al. 2001, Scherzinger 1996, Peterken 1993).

Preservation of soils

Forest management has to observe and maintain the natural site qualities in both the short and long term. All soils from "productive" to "extreme" must be preserved while avoiding the melioration and levelling of soils. The deposition of acids and nitrogen must be reduced and compensation measures, such as the liming of forest floors, only carefully applied. Diversity of tree species ensures a diversity of soil life, and thus may improve soil fertility.

Naturalness and originality

Naturalness cannot be observed and measured directly as it is a theoretical conceptual construction. Naturalness must, however, be highly valued, particularly in production forests. Near natural forestry reduces human impacts and contributes to preserving site-adapted species and biocoenoses. In the near future the definition and degree of naturalness itself will change due to changes in CO_2-emissions, climate and modified atmospheric chemistry. The rapid degradation of naturalness in ecosystems formed by long-lived species, such as trees, reduces its value for nature conservation. Instead of naturalness, the originality criterion must therefore be given an increasingly higher weight. Originality is a mental construction of a natural state (1) under similar climatic and biocoenotic conditions compared to today, but (2) before human impact altered the ecosystems (Reif 2000, Peterken 1993). For nature conservation, the preservation of original ecosystem requisites is a highly ranked aim, including the preservation of terminal and decaying phases of old-growth forests.

In forest reserves and nature conservation areas it will be particularly necessary and challenging to preserve the native species and the genetic diversity of their populations. Old forests must be given a particularly high value in open landscapes because less mobile species will only survive in such systems. Anthropogenic destruction of the original characteristics of habitats is critical because the consequences will only be realised in the long term and may be irreversible.

Rarity and endangerment

Certain site conditions, species and biocoenoses are, and always have been, rare. To increase their frequency should not be the aim of nature conservation. The endangerment of populations must be recognised and analysed, and measures to preserve endangered species and habitats should be undertaken. Critically endangered are forest species of old growth phases and of extreme sites. This includes forests with valuable broadleaved tree species on sites with naturally dynamic vegetation, e.g. on unstable slopes along rivers and streams. Rare species of these habitats can become endangered very quickly. For this reason rare species need particular care and their sites effective protection.

Diversity

Diversity exists on different hierarchic levels; that of (1) the species and their populations (genetic diversity), (2) communities and their species richness and structural diversity and (3) the landscapes with their community diversity. Production forests can exhibit relatively high diversity, as can be shown by comparing managed and unmanaged forests. Increased species numbers often result from human impact, e.g. through planting new species, meliorating extreme sites or liming forests on acidic sites. These activities favour nitrophytic species of disturbed sites or widespread, rather than specialised, species.

High diversity can be the result of historic forest uses, often leading to an increase in oak (*Quercus*) and other light-demanding species at the expense of the shade-tolerant beech (*Fagus sylvatica*). Following recent land use changes, beech has reinvaded many sites from which it was once eliminated, e.g. sand dunes and clay soils. The consequence of this will be a decline in light-demanding trees such as *Quercus*, pine (*Pinus sylvestris*) and associated species, including many specialists of light and nutrient-poor sites.

The site-dependent diversity of species and forest structures should be preserved and maintained. The criterion diversity must be viewed in relation to the high values of the originality and naturalness, and rarity and endangerment criteria. An increase in species diversity will in many cases indicate a reduction of the value of nature conservation because other criteria are negatively affected.

5.2.3 Nature preservation and management of valuable broadleaved forests

Nature preservation in managed valuable broadleaved forests has to include aspects of protection of the abiotic resources and of species and their habitats on different scales (Table 5.1).

Abiotic resources

The litter of valuable broadleaved tree species decomposes rapidly and deterioration of the base saturation and buffering capacity of the topsoil does not occur. However, the use of herbicides is recommended in France after planting young stands (Société Forestiére de Franche-Comté 1998, Armand 1995).

Site conservation problems may arise through melioration, e.g. when swamp forests are drained, and in cases where rare and endangered forest communities are converted to stands of valuable broadleaved tree species. On drained soils peat decomposition and increased nitrification favour widespread nitrophytic species while the endangered original swamp forest species disappear.

- In the Rhine river valley most alder swamp forests have been drained over the last decades. Following peat decomposition and nutrient release, valuable broadleaved forests replace the "relic" *Alnus glutinosa* and the specialised associated flora (Pretzel et al. 1997; Figure 5.29).
- In north east Germany valuable broadleaved tree species are recommended for wetland forest sites where they will replace *Alnus glutinosa* (Ministerium für Landwirtschaft und Naturschutz des Landes Mecklenburg-Vorpommern 1995).
- Alder (*Alnus glutinosa*) is frequently planted on clearings as a nurse tree for oak (*Quercus robur*). Through symbiotic nitrogen fixation alder increases the nitrogen content in the soil, promotes nitrophytes and contributes to more rapid humus mineralisation.

Table 5.1. Influence of planting or managing forests with valuable broadleaved tree species (VBTs) and the effects on nature preservation, regarding the genetic, species, stand and landscape levels, and including three important criteria of evaluation.

Levels \ Criteria	Originality and naturalness	Rarity and endangerment	Diversity
Landscape and stand	• Forestry: Conversion of near-natural tree compositions, e.g. *Fagus sylvatica*-dominated forests, towards stands with VBTs reduces originality and naturalness. • Afforestation of open land may increase the naturalness with time, e.g., planting VBTs on base-rich soils.	• Afforestations in endangered cultural habitats, e.g. low fertility grasslands, reduces the landscape diversity and endangers rare open-land species. On base-rich soils, VBTs are dominant or common components in afforestations.	
Stand	• Forestry destroys old-growth natural forests and their specific landscape-forming potential, e.g., when replacing a previous *Carpinus-Quercus*-forest by a VBT stand. • Silviculture: Pure stands of VBTs are evenaged, evensized and harvested relatively early (decrease of naturalness). • Planting of VBTs may support melioration (drainage) of soil (destruction of originality). • The understory of VBT plantings often consists of widespread "ruderals" (= herbaceous annual or biennial species colonizing disturbed sites) and nitrophytic species (low frequency or absence of specific forest floor species; low values in naturalness, rarity, endangerment).		Small- to medium-scale forestry increases the landscape diversity.
Species		• Silvicultural treatments may promote or reduce rare and endangered species, depending on the situation. Particularly, species of old growth forests are eliminated.	• Silvicultural measures such as tree planting or inducing an "intermediate disturbance" increase the species diversity.
Genetic	Endangerment or destruction of the original genetic variability, genetic composition and adaptability of tree species by forestry practices: • production of transgenic plants, • plant breeding, • thinning: Elimination of unwanted morphotypes.		• Forestry: Planting material from other regions ("not autochthonous") may increase genetic diversity locally, but will contribute to levelling and thus the reduction/destruction of genetic variability within the area of the species. • Forestry: Tree breeding and transgenic trees may result in a new, man-made genetic diversity.

Figure 5.28. Drainage of a previous alder (*Alnus glutinosa*) swamp forest where the decrease of the ground water level mineralized the peat. The stilt-rooted alder trees were able to survive some decades but will be replaced by maple-ash-plantings in the next stand generation. – Wasenweiler Ried, SW Germany, 1996.

As these examples underline, the cultivation of valuable broadleaved tree species is an environmentally friendly silvicultural technique. However, in some cases the melioration of sites precedes cultivation and endangers rare forest species and communities.

Preservation of species and habitats

Three hierarchic levels for the protection of forest species and habitats exist (Larsson et al. 2001); namely: (1) the genetic diversity of populations (species level), (2) the preservation of sites, species composition and stand structures (community level) and (3) the preservation of landscapes; all of which should be discussed separately. Naturalness and originality, rarity and endangerment, and diversity are the criteria required for an evaluation of management measures in production forests (Reif et al. 2001).

Genetic level

In the long term the conservation of species has to coincide with species adaptation to existing site conditions and their genetic diversity and adaptability to environmental changes. Within the environmental range of a tree species "autochthony" is a principal indicator which means "native" or "naturalised for an unknown length of time". Autochthonous populations have adapted during the course of evolution to specific site

Figure 5.29. Old, twisted *Acer pseudoplatanus* in a forest reserve. In production forests these morphotypes will not be able to survive for a long time. – Falkenstein, Bayerischer Wald/Bavaria, 1987.

conditions. Populations on extreme sites and near the border of their distribution area, in particular, will contribute to adaptability and evolution in the future.

To date, the selection of fast-growing, productive and stable varieties has been the main aim of forest tree breeding practices in several European countries. The genetic properties of unwanted morphotypes of trees are eliminated from production forests (Figure 5.30). The consequence is a reduction of the genetic potential of the naturally evolved native populations and partial replacement by a new, "forestry-adapted" genetic diversity. This may reduce the future capacity for adaptation under changing environmental conditions.

The use of valuable broadleaved tree species seeds and juvenile plants is regulated by law in several European countries. The use of regional provenances allows at least for the adaptation to climate. Concerns for nature conservation remain however:

- Seed trees must have straight boles and fine side branches. The same is true when extracting and planting seedlings and saplings taken from the wild (Ministerium für Landwirtschaft und Naturschutz des Landes Mecklenburg-Vorpommern 1995). The consequence of such genetic selection of forest trees is the reduction of genetic diversity.
- Economically unproductive, although site-adapted, genotypes are not promoted in forestry and unwanted ecotypes, races or provenances will be absent from newly established production forests (Bund-Länder-Arbeitsgruppe "Erhaltung Forstlicher Genressourcen" 1989).
- The use and extraction of young trees from the few remaining valuable broadleaved mother trees in the wild is rejected in Bavaria because the genetic

diversity of isolated trees is regarded as low. This ignores the fact that the alternative option of obtaining planting material from a relatively large provenance area contributes to genetic levelling in forested landscapes (Bayerisches Staatsministerium für Ernährung, Landwirtschaft und Forsten 1999).

Silvicultural measures for tree mixture regulation and thinning represent genetic selection and induce genetic drifts in tree populations. For example, longevity will not be a selective factor in production forests because the trees will be harvested long before reaching their natural mortality.

Forestry induces shifts of genetic composition and losses of genetic diversity of forest trees, including valuable broadleaved tree species. This is also true for populations and stands derived from natural regeneration. Compromises between forestry and nature conservation can be found when valuable broadleaf hardwood stands are managed with natural tree species regeneration, provided the existing stands are genetically diverse and from autochthonous provenances.

Species level

During the course of central European forest history the shade-tolerant tree species beech (*Fagus sylvatica*), fir (*Abies alba*), but also other trees like hornbeam (*Carpinus betulus*) and oak (*Quercus* spp.) have lost much of their original area. On the other hand, in milder climates tree genera such as oak (*Quercus*), hornbeam (*Carpinus betulus*) and associated species such as *Acer campestre*, *Prunus avium* or *Sorbus*-species have increased in abundance at the expense of beech. Under less favourable climates and on acidic soils, forest clearing and subsequent afforestation has resulted in the reduction of deciduous trees, with conifers dominating much of today's landscape.

Valuable tree species of the genera *Sorbus* and *Acer* have become rare and are endangered in some regions (Jedicke 1997).

- *Sorbus torminalis* is recommended for base-rich, wet to medium sites in north east Germany. Wild cherry (*Prunus avium*) should be planted in gaps of regenerating beech (*Fagus sylvatica*) stands on loamy soil (Ministerium für Landwirtschaft und Naturschutz des Landes Mecklenburg-Vorpommern 1995).

Depending on the species, site and region, forestry can promote or endanger valuable broadleaved trees. Planting of *Sorbus torminalis* and *S. domestica* and the promotion of naturally regenerated juveniles may contribute to the preservation of these species.

Stand and landscape level

The natural vegetation of central Europe includes several forest types consisting of or containing valuable broadleaved trees (Stortelder et al. 1999, Mucina et al. 1993, Schmider et al. 1993, Oberdorfer 1992, Rodwell 1991, Mayer 1984). Valuable broadleaved tree species can dominate naturally on unstable slopes or montane alluvial soils in association with rare, specialised species. The corresponding forest communities are protected by law in several regions and countries of Europe. Commercial use of such forests would modify the sites and reduce the species and genetic diversity of trees and their associated species. On the other hand, valuable broadleaved trees have increased on a range of sites and management types in production forests.

a) Production forests dominated by valuable broadleaved trees
Production forests dominated by valuable broadleaved trees have often been established on nutrient-rich soils naturally dominated by beech (*Fagus sylvatica*), e.g. the drained soils of former wetland or alluvial sites. Due to the high nutrient supply, widespread nitrophytic species cover the forest soil instead of typical forest floor species. The canopies of the planted stands often contain species which would not naturally be competitive on these sites. The resulting stands therefore have a reduced degree of naturalness (Landesforstverwaltung Baden-Württemberg 1999).

- In France the planting of valuable broadleaved trees in groups or stands of up to two hectares is recommended (Société Forestiére de Franche-Comté 1998). Juveniles are planted systematically in rows with 1100 light saplings of ash (*Fraxinus excelsior*) per hectare (Société Forestiére de Franche-Comté 1998) or 200 to 600 saplings ("Heister") per hectare (Armand 1995). The plantings are to be treated every five to seven years up to an age of 35 years, with longer intervals thereafter. Ultimately ca. 50 to 70 ash and 70 to 90 cherry and maple trees will be harvested per hectare (Société Forestiére de Franche-Comté 1998).
- In Germany even aged stands dominated by valuable broadleaved trees have been, and continue to be, planted at the expense of beech. Their degree of naturalness is medium to low with rare and endangered species missing completely. Only recently have concepts to convert such stands to production forests of uneven sizes and ages been developed (Landesforstverwaltung Baden-Württemberg 1999).

On the moist colluvium at the bottom of slopes or adjacent to creeks unstable conifer plantations have recently been replaced by valuable broadleaved species including alder (*Alnus glutinosa*) (Bayerisches Staatsministerium für Ernährung, Landwirtschaft und Forsten 1999). Naturalness and diversity increase as a result and the associated riparian biocoenoses may become restored with time.

Other valuable broadleaved forest stands originate from succession on abandoned agricultural land, or from the afforestation of previous agricultural land or grassland; in many cases through the subsidised conversion of moist meadows (Figure 5.31).

- The CO_2 fixation may also be an aim of afforestation, but its effect is discussed controversially in relation to the large amount of the total CO_2 emissions.
- In recent years many afforestation projects have been carried out with the intended function of compensating environmental losses, e.g. through urban and industrial areas, after building new roads etc. In many cases these losses cannot be compensated because valuable open-land habitats have been destroyed (Reif and Nickel 2000).

Figure 5.30. Afforestation of an previous wet grassland with Fraxinus excelsior. The naturalness of the tree species composition will increase, but valuable openland habitats disappear. – Hollfeld, northern Bavaria, 2002.

b) Mixed production forests with a proportion of valuable broadleaved trees

Mixed production forests with a valuable broadleaved tree component should preferably be established on nutrient-rich, moist or dry sites of beech (*Fagus sylvatica*) forest. Management tends to increase the amount of valuable broadleaved trees, resulting in increased structural and tree species diversity and decreased naturalness.

- Valuable broadleaved species appear as single trees or small groups in oak-dominated and beech-broadleaf mixed forests ("Buchen-Laubbaum-Mischwald"; Landesforstverwaltung Baden-Württemberg 1999).
- Approximately one hundred cherry (*Prunus avium*) trees per hectare may be a temporal component of beech forests (Bayerisches Staatsministerium für Ernährung, Landwirtschaft und Forsten 1999; Centre Regional de la Propriete Forestiére, without year).
- Small-leaved lime (*Tilia cordata*) is promoted as a serving tree which prevents the stem buds of oaks from sprouting (Ministerium für Landwirtschaft und Naturschutz des Landes Mecklenburg-Vorpommern 1995).
- Valuable broadleaved trees in small stands, mixed with and in the understory of other trees, should be stimulated if they are of good quality. Their proportion should be increased from 2.6 to 7.5 % of the forested area for the purposes of increasing biological diversity and reducing biotic and abiotic risks (Ministerium für Landwirtschaft und Naturschutz des Landes Mecklenburg-Vorpommern 1995).
- Maple (*Acer pseudoplatanus, A. platanoides*) are increasingly being planted on sites where it is naturally absent. These species are tolerant of high nitrogen content (Ministerium für Landwirtschaft und Naturschutz des Landes Mecklenburg-Vorpommern 1995).

- Rapid growth in the early phases of development (between 5 and 25 years) leads to dense pole stands with closed canopies and the onset of natural pruning (Bayerisches Staatsministerium für Ernährung, Landwirtschaft und Forsten 1999). Most vascular plants in the understory cannot survive during this phase.
- The regeneration of valuable broadleaved species should be promoted by applying shelterwood and selection forest systems, after which groups should be formed (Bayerisches Staatsministerium für Ernährung, Landwirtschaft und Forsten 1999).

The evaluation of valuable broadleaved species production forests depends on the previous vegetation, the site and subsequent stand treatments. Mixed stands can exhibit greater tree species diversity than beech dominated stands. Harvesting of the individual trees after ca. 70 to 100 years (Bayerisches Staatsministerium für Ernährung, Landwirtschaft und Forsten 1999) hinders the development of large-timbered stems and impacts fauna that depend on old trees for survival. A network of old stands, snags and forest reserves is necessary to maintain a sufficient degree of naturalness and protect the typical forest species in the forested landscape (Larsson et al. 2001).

References

Armand, G. (ed.) 1995. Feuillus précieux. Conduite des plantantions en ambiance forestiére. IDF Paris. 112 p.

Bayerisches Staatsministerium für Ernährung, Landwirtschaft und Forsten (ed.) 1999. Waldbaurichtlinien. Anlage: Pflegegrundsätze für Edellaubbaumarten und Schwarzerle. München. 20 p.

Bund-Länder-Arbeitsgruppe „Erhaltung Forstlicher Genressourcen“ 1989. Konzept zur Erhaltung forstlicher Genressourcen in der Bundesrepublik Deutschland. Forst u. Holz 44: 379–404.

Centre Regional de la Propriete Forestiére (ed.). Qualité du Bois et Sylviculture de l,Erable Sycomore. Amiens. 16 p.

Centre Regional de la Propriete Forestiére (ed.). Qualité du Bois et Sylviculture du Fréne. Amiens. 17 p.

Ellenberg 1996. Vegetation Mitteleuropas mit den Alpen. Ulmer, Stuttgart.

Jedicke, E. (ed.) 1997. Die Roten Listen. Gefährdete Pflanzen, Tiere, Pflanzengesellschaften und Biotoptypen in Bund und Ländern. Ulmer, Stuttgart. 581 p.

Landesforstverwaltung Baden-Württemberg (ed.) 1999. Richtlinie landesweiter Waldentwicklungstypen. Stuttgart. 54 p.

Larsson, T.-B. (coord.) 2001. Biodiversity Evaluation Tools for European Forests. Ecol. Bull. 50: 237 p.

Lecomte, H., Florkin, P. and Thirion, M. 1997. L´Inventaire des massifs forestiers de la Wallonie: Apercu global de la situation en 1996. Jambes. 42 p.

Mayer, H. 1984. Waldbau. 3. Aufl., G. Fischer, Stuttgart, New York. 514 p.

Ministerium für Landwirtschaft und Naturschutz des Landes Mecklenburg-Vorpommern (ed.) 1995. Edellaubholz-Programm für den Landeswald von Mecklenburg-Vorpommern. Manuscipt. 5 p + annexes.

Mucina, L., Grabherr, G. and Wallnöfer, S. (eds.) 1993. Die Pflanzengesellschaften Österreichs. Teil III. Wälder und Gebüsche. G. Fischer, Jena, Stuttgart, New York. 353 p.
Oberdorfer 1992. Süddeutsche Pflanzengesellschaften. Teil IV: Wälder und Gebüsche. Tabellenband 580 S., Textband 282 S. G. Fischer, Jena, Stuttgart, New York.
Peterken, G.F. 1993. Woodland Conservation and Management. 2. edition. Chapman & Hall, London. 374 p.
Pretzell, D., Knör, E.-M. and Reif, A. 1997. Degradation von Erlenbruchwäldern in der Oberrheinebene. Verh. Ges. Ökol. 27: 435–440.
Reif, A. 1997. Zielkonflikte im Naturschutz – Kontroversen und ihre Ursachen am Beispiel der Diskussion um die oberrheinische Trockenaue bei Breisach. Naturschutz und Landschaftsplanung 29: 101–107.
Reif, A. 2000. Das naturschutzfachliche Kriterium der Naturnähe und seine Bedeutung für die Waldwirtschaft. Z. Ökol. U. Naturschutz 8: 239–250.
Reif, A., Knoerzer, D., Coch, T. and Suchant, R. 2001. Landschaftspflege in verschiedenen Lebensräumen. XIII-7.1 Wald. In: Konold, W., Böcker, R. and Hampicke, U. (eds.). Handbuch Naturschutz und Landschaftspflege, 4. Erg.Lfg. 3/01. Ecomed-Verlag, Landsberg. 88 p.
Reif, A. and Nickel, E. 2000. Pflanzungen von Gehölzen und „Begrünung" – Ausgleich oder Eingriff in Natur und Landschaft? Naturschutz und Landschaftsplanung 32: 299–308.
Rodwell, J. 1991. British Plant Communities. I. Woodlands and Scrub. Cambridge University Press, Cambridge.
Scherzinger, W. 1996. Naturschutz im Wald. Qualitätsziele einer dynamischen Waldentwicklung. Ulmer, Stuttgart. 447 p.
Schmider, P., Küper, M., Tschander, B. and Käser, B. 1993. Die Waldstandorte im Kanton Zürich. Waldgesellschaften, Waldbau, Naturkunde. VdF, Zürich. 287 p.
Société Forestiére de Franche-Comté (ed.) 1998. Les Feuillus Précieux en Franche-Comté. Thise. 28 p.
Stortelder, A.F.H., Schaminée, J.H.J. and Hommel, P.W.F.M. (eds) 1999. De Vegetatie van Nederland. 5. Ruigten, Struwelen, Bossen. Opulus Press, Uppsala, Leiden. 376 p.
Volk, H. 2001. Auewaldforschung am Rhein – welche Wälder sind auewaldtypisch? Natur u. Landschaft 76: 520–528.

5.3 Do Species Matter? Valuable Broadleaves as an Object of Public Perception and Policy

Ulrich Schraml and Karl-Reinhard Volz
Institute of Forest and Environmental Policy, University of Freiburg, Germany

Abstract

Changing public awareness concerning environmental issues is often expected to be a driving force in the planning and management of forests. Experts generally regard broadleaved trees and particularly valuable broadleaved species as vital elements of multi-purpose forestry both in rural areas and in urban settings. This chapter examines the importance of species in the individual's and the public's perception of forests. It therefore gives an overview of theories explaining forest preferences and describes the crucial role of valuable broadleaves in culture and politics. Finally, this chapter investigates their relevance in people's minds by reviewing surveys that were published throughout Europe.

Keywords: broadleaved trees, myths, culture, forest policy, opinion polls

5.3.1 Introduction

The selection of tree species is among the most important decisions in any forestry operation. This decision becomes especially acute when the species composition of a given forest no longer meets the requirements of the owner or when a new forest planting is newly established. However, throughout history the selection of tree species has been discussed and influenced not only by the owners but also by a number of other interested parties. Various forest users and their political representatives have become aware of the fact that their chances of realising specific interests or demands are directly affected by the existing tree species in an area, and have thus begun to influence forest owners and foresters with regard to this topic. In European countries this influence has led to the development of an extensive set of political instruments which are now also affected by the international forest policy.

Compared to the interests of the political actors, the existence of a public perception and evaluation of individual forest types and tree species by laymen is much less apparent. The Swiss sociologist Wild-Eck (2002) offers a rather pessimistic view: "The forest rarely evokes everyday thoughts in most people. Its existence is taken for granted and it will only stimulate an extensive thought process once it is endangered or altogether absent." This view is in stark contrast to the numerous descriptions of the cultural value of forests in which valuable broadleaf species such as ash or elm in particular play a central role (Mayer-Gampe 1999, Harrison 1992).

Explanations for this contradictory assessment of the perceived importance of nature and its elements are given in a steadily growing number of publications. The current scientific debates on the social relevance of nature often focus on social constructionist approaches to explain intergroup differences. "Environmental attitudes, feelings, actions, and perceptions are not simply free-floating; rather they need to be seen in context" (Marsden et al. 2003). In this chapter we further examine the perceptions and preferences of the specialists and the average citizen on forest types by reviewing existing empirical studies. In detail, we will first offer some theories, concepts and ongoing processes from the field of social sciences, psychology, culture and politics which may explain the responses elicited by visual or cognitive encounters with valuable broadleaved trees and forests. Secondly, we present and review surveys from different regions in Europe that take the different social contexts of people into consideration. The emphasis is on valuable broadleaved trees although the referred to studies often do not clearly distinguish species and therefore we sometimes focus on broadleaves in general.

5.3.2 Explaining forest preferences

Forest assessment

Any assessment of forests is based on the perception of physical elements by an individual or a group of people. Whether the availability and distribution of these physical elements or the emotional and cognitive requirements of the observer are considered as the dominant elements of the evaluation process represents a highly important assumption in regard to the expected results. Therefore, any study of the perception of forests by experts, political actors or the public implies a decision favoured by the use of a particular assessment model (Lee 2001).

Assessment models that are based on knowledge of ecology and economy or on formal aesthetic principles focus strongly on the physical elements of a forest. They represent the typical expert approach and aid primarily in the evaluation of forest types for planning and management purposes. Therefore 'objective' information about forest quality is an expected result. By virtue of their training and their heightened sensitivity, experts are able to draw fine distinctions and define the proportion of tree species in a forest as a crucial element.

Models for the assessment of forests by the public generally emphasize the subjective meaning of the environment for the individual. The personal feelings, preferences, interpretations and associations of an observer are strongly influenced by his socialisation. Due to the large variety of social and cultural contexts in modern society, a comparably high level of variety in the significance of the forest for individuals or different groups can be expected. On the other hand, owing to the common heritage of humans in terms of evolutionary and cultural history, certain similarities in people's reactions to different tree shapes are assumed (Sommer 1997, Sommer and Summit 1996).

Biological and social factors

Leading authors dealing with environmental perception and preferences refer to evolutionary theory, natural selection or habitat patterns to explain landscape perception and preferences (Orians and Heerwagen 1992, Kaplan 1987, Orians 1986, Wilson 1984, Appleton 1975, Dewey 1958). Several of these theories are also helpful in explaining forest preferences.

The observed complexity appears to be a crucial element in forests that are positively evaluated. Diverse habitats with many amenities offer stable ecosystems, relative safety and better living conditions for man. Thus, a preference for landscapes with a high level of complexity is adaptive in an evolutionary sense (Kaplan 1992).

This explanation may at least be valid for people who lack experiences that could have generated individual or group-specific preferences. The preferences of those who were able to gain particular experiences in the surroundings in which they grew up are thought to be modified according to local landscape conditions. Some authors who use evolutionary theories postulate a learning process in individuals who recognise their parents' habitat as optimal for raising their own offspring (Balling and Falk 1982).

Aside from the subconscious preferences that may have developed as a result of natural selection in a specific regional culture, cognition is an important factor in the evaluation of landscapes (O´Leary and McCormack 1995). For example, a preference for diverse forests may not only be an advantage in an evolutionary sense but may also be fostered by the widespread knowledge of the relationship between diversity and ecosystem stability. Already the perception of forest is filtered by normative ideas, conceptual understandings or associations with what is seen. In particular, the acceptability of management impact on forests is influenced by the information provided (Kearney et al. 1999, Ribe 1989).

In summarizing these variables, various authors stress that "belonging to a particular social group influences considerably the individual's perceptions of nature and forests". The meaning that different forests and forestry practices may have to an individual varies in relation to the contextual setting (Schmithüsen and Wild-Eck 2000). The 'group' (Buchanan et al. 1981), 'stakeholders' (Piussi and Pettenella 2000) or the 'milieu' (Braun 2000) represent the social context responsible for influencing the perceptions.

Adaptation of forest perception and preferences

Adaptation processes serve as another important approach in explaining forest assessment. One typical criterion for the characterisation of forests is their steadiness and long-term continuity. Ongoing processes that could cause changes in the mixture of trees, such as growth processes or regeneration, are rarely observed by non-professionals unless they are explained in detail. Therefore they are subject to special conditions if they are to serve as objects of perception and preferences.

Several psychological studies show that stimuli that become constant provoke a decreasing level of reaction or response in the observer. Psychology has investigated this phenomenon and has termed it adaptation. It can be explained either cognitively or psychologically. Psychological explanations suggest that the receptors, as main elements of the human perception system, react less frequently to repeated stimuli. Cognitive explanations refer to the observation that a stimulus, once it has been identified as something harmless, will not demand the same attention upon regular presentation. The latter explanation reflects a response bias rather than a perceptual shift (Hook 2002).

In this sense, adaptation may occur in the perception of forests. Even the perception of forest types considered by experts to be of lesser quality (e.g. forests dominated by conifers) may undergo adaptation in order to avoid cognitive dissonance. On the subject of appreciation of tree species distribution, it was postulated as early as the early 1970s that the indifference of the population toward the mixture of tree species increases the less the species composition deviates from the composition of the forests which people visit for recreation (Heeg 1971).

The phenomenon of "adaptive preference" was suggested by Elster (1983) in the context of supplementing "Rational Choice" approaches in economic theory. He metaphorically explained this adaptation process with the legendary reaction of the fox who rejected the grapes as sour once he realised they were unavailable. Thus the term "adaptive preference" describes the phenomenon of individuals adapting their preferences to their circumstances, especially in view of perceived constraints. People prefer certain goods because alternatives are not available and dislike other goods that are not available. This also lends significance to the probability of a shift in preferences. Once adapted, preferences can be reversed. Any change in constraints or circumstances may cause other shifts in preferences.

5.3.3 Significance of valuable broadleaves

Valuable broadleaves in a social and cultural perspective

The development of preferences for valuable broadleaves first requires the existence of such trees, either within the real, personal surroundings of an individual or as part of the knowledge imparted through personal communication or the media. The relative scarcity of valuable broadleaves impedes the initial formation of preferences for these species by way of adaptation. However, personal experiences with the trees and the use of their wood can be developed even for rare species if they are specifically sought out.

While the use of wood in crafts is nowadays restricted to a rather limited group of people, the special experience value of valuable broadleaf species can lead to positive encounters for many forest visitors. Maple, walnut, lime and cherry contribute to this through the use of their fruits and leaves as toys, medicine, food or decorative items. Furthermore, the lime tree holds a proverbial place at the centre of society as it was often planted around farms, at crossroads and on the village greens. According to a common literary theme, lovers meet under its branches and the social life of the villagers takes place in the shade of the linden tree.

Another venue for a closer approach is offered through mythology and religion (Philpot 1897). One classic example from the tradition of Central and Northern Europe is the ash tree "Yggdrasil", the broadleaf tree which, according to Germanic beliefs, holds together the universe and at the same time represents the connection between gods, humans and the realm of the dead. The lime tree, dedicated to the goddess Freya, has also achieved cultic significance. There is also the dominant role which oak has played in many European countries even since pre-Christian times. The etymology of the word "oak" shows common Indo-Germanic roots (*Eiche*, *être*) and it was equally revered by Greeks, Germanic tribes and Slavs (Hilger 1956). The successful incorporation of tree cults into the Christian belief becomes apparent in quantitative studies of place names. In France and Germany a surprising number of towns and villages whose names are derived from tree names serve as places of pilgrimage (Marzell 1929). The variety of revered broadleaf trees is further documented in a survey conducted in 1854 in the French department of Oise. Among the 253 "*arbres vénérés*" (venerated trees) listed in the survey were 74 elm, 27 oak, 15 walnut, 14 beech and 14 lime trees (Sebillot 1906 in Marzell 1929).

A third traditional approach to valuable broadleaved trees is through symbolism based on their impressive size and age. Again, at least in the temperate forest belt, it is mainly the broadleaf species which, as solitary trees, can reach such "mighty" proportions. Thousand-year-old oaks, beeches and lime trees have been documented from various regions in Europe at least since the Romantic period (Deutsches Baumarchiv 2004, Kühn et al. 2002, Klein 1908, Großh. Ministerium 1904). These "mighty" trees reach impressive stature through their overall shape and height, but in most cases it is their unusual diameter that stands out. This is typical for the oak tree which in several countries has found its way into the national symbolism (see Keith 1983 about "English Oak", Lehmann 1999 about "German Oak"). Maple and lime trees can also reach dimensions which render them symbols of strength and durability.

Another approach to the development of social preferences may be found through literature and poetry. The forest is also a literary construct. It can therefore be assumed that society's preference for valuable broadleaf trees may be traced back to their occurrence in the national literature. This can be confirmed in the so-called "Freiburger Anthologie" (Freiburg Anthology) (2002) for 1282 German poems from the period between 1720 and 1900. Among the tree species mentioned here, the oak indeed ranks first over lime and beech followed some rankings lower by elm, alder, birch and cherry. It appears that the maple tree has few friends among the German poets as it is not mentioned at all. Conifers clearly rank below the broadleaves. The truly remarkable result of this analysis, however, is the fact that in all analysed poems the term "tree" was used about three times as often and the term "forest" about eight times as often as the name of the most commonly mentioned tree species, the oak. It appears the poets of the 18th and 19th century too were specialists of language, not botany, and did not assign a special role to the valuable broadleaves.

This gives rise to certain scepticism as to the traditional significance of valuable broadleaves in urban societies. Both the knowledge of the properties of wood and fruits as well as the specific experience value of individual tree species are increasingly lost. In everyday life the conscious use of particular parts of valuable broadleaf trees is restricted to individual social groups. Furthermore, valuable broadleaves stand in heavy competition with other tree species for public awareness. Due to the Christmas tradition, spruce and fir are often the first and sometimes only tree species which people in modern society learns to know and identify. There are also several other species other than valuable broadleaves that are revered as cult trees or mentioned in literature and fairy tales. Therefore the societal preference for valuable broadleaves can only be partially explained through usage, mythology, symbolism and literature.

5.3.4 Valuable broadleaves in forest policy

Forest policy topics

Tree species may also be brought to the public's awareness if they become the subject of conflicts and subsequent media coverage. Tree species selection was among the disputed topics during the environmental forestry conflicts in Europe in recent decades (Hellström and Rytilä 1998, Hellström 1996, Hellström and Welp 1996). The often described changing role of forestry in urbanised societies supports the inclusion of social and ecological demands into wood production and urban greening (Konijnendijk et al. 2005, Schraml 2003, Kennedy et al. 2001, Konijnendijk 1999, Miller 1988). Aside from this shift in values, in the 1990s the focus of the discussion shifted increasingly from local and regional problems to those at the global scale. As a necessary result, the terms used to define policies are becoming increasingly abstract so that little mention of tree species or species groups can be expected in international documents.

One example from the international forest policy is the demand for increased biodiversity in forests. Already in 1992 the United Nations' Conference on Environment and Development ratified the so-called "Forest Principles" with the goal "to contribute to the management, conservation and sustainable development and to provide for their multiple and complementary functions and uses" (UNCED 1992). To that end, "the vital role of all types of forests" in maintaining ecological processes, protecting ecosystems, watersheds, freshwater resources and, at the least, biodiversity was stressed. Efforts to "green the world", namely through afforestation and forest conservation, were postulated.

On the international and European as well as the national level, several efforts were made to implement these principles and the contents of Agenda 21 (MCPFE 2001, COM 1998). In 1998 the Third Ministerial Conference on Protection of Forests in Europe (MCPFE) addressed the forest-related decisions and agreements on the international level through UNCED, the United Nations' General Assembly Special Session or the United Nations' Conventions on Biological Diversity and Climate Change. MCPFE (2003) again stressed the economic viability as well as social and cultural dimensions of sustainable forest management in Europe.

When dealing with forest productivity, cultural values, recreation, biological diversity and water conservation, both the Pan-European Criteria and Indicators and the Guidelines for Sustainable Forest Management frequently touch upon the role of tree species considered native, rare, attractive and of high cultural value. In most European regions broadleaves in general and valuable broadleaves in particular fulfil these expectations in every respect; nevertheless, the papers are too abstract to focus on forest types or species.

The European Union is trying to implement the international framework with its Forestry Strategy. According to the restricted responsibility of the union, the most important forest-related measures are in the fields of rural development (agriculture), environmental issues and international markets (COM 1998).

Afforestation policy

The idea of "greening the world" founded a European action in some of the forestry measures in agriculture. As one of the accompanying measures to the 1992 Common Agricultural Policy (CAP) reform, Regulation 2080/92 aimed at the afforestation of agricultural land and improvement of forest resources in the European Community (EC). In the member nations the regulation has been met with ongoing changes in national policies in this field. Therefore, the implementation on a national level and in federal systems on a regional level has led to a high diversity of policies within Europe. In particular, the EC's objective to stimulate the plantation of broadleaves or mixed-species plantations encountered heterogeneous national circumstances, political goals and capacities. A later evaluation report covering the consequences of afforestation in eight member nations revealed that the afforested area of one million hectares in the EC, resulting from Regulation 2080, are dominated by broadleaves while some areas are covered with mixed stands. With the exception of Ireland, the national programmes have all encouraged the use of broadleaved trees. In some countries (e.g. Austria, Germany and Finland) the implementation of the aid scheme to create a greater diversity in existing forest stands contributed additional success (Institute for Forestry Development 2001).

In accordance with the different starting conditions and the differences in national forestry policies, as described in Table 5.2, the changes in species composition through afforestation appear very heterogeneous (Table 5.3). The different countries have chosen very different options regarding the overall changes of the total forest area as well as changes in tree species composition. This impression is further confirmed when considering the individual species that were used. Mid-term broadleaves such as black walnut (*Juglans nigra*), ash (*Fraxinus* sp.) and wild cherry (*Prunus avium*) occupy 13.1% of the afforested area in the eight countries shown in Table 5.3. Ultimately this reflects the special political efforts undertaken by the individual countries to plant "noble" species. In France walnuts (*Juglans* sp.) show a considerable percentage with 6% of the afforested area while in Italy the mixtures of valuable and semi-valuable broadleaves constitute an estimated 70% of the new forests. Unfortunately most countries failed to keep precise records of the species used in the plantations and therefore only limited statements on this level are possible. In all of Europe the percentage of valuable broadleaves in the documented afforested areas lies between 2.4% (*Juglans* sp.), 1.8% (*Prunus* sp.), 0.4% (*Fraxinus* sp.) and 0.0% (*Acer* sp.).

Conversion of conifer stands

In some Central European countries the afforested area is quite small. In Germany, for example, only 27,045 ha were newly afforested. However, a large proportion of

Table 5.2. Examples of the implementation of Regulation 2080/92 in four countries (Institute for Forestry Development 2001).

	Selected national objectives	Characteristics of measures
Germany	Increase stability and longevity of forests, achieve better vertical structure, encourage a mixture of species.	Improvement area larger than afforestation area, long-term broadleaves like oak (38%) and beech (19%) used as main species for planting, cherry and alder about 5% each and others for a total of approximately 10 different species.
Denmark	Enlarge forested area, increase rural development, recreation, water quality and shelter belts.	Planting of mainly long-term broadleaves (oak as dominant species) in peri-urban areas, high efforts to foster diversity even in individual stands, minimum 14 species used.
Italy	Increase rural development, create income for farmers, increase high added value products and water quality.	High percentage of valuable broadleaves such as *Juglans* sp., *Prunus avium*, poplars, *Alnus* sp., innovative plantations along water courses.
Ireland	Enlarge forested area, increase timber production.	High involvement of public forestry bodies, no improvement areas, 84% conifers (*Picea sitchensis*).

the improvements (101,000 ha) financed under Regulation 2080 aimed at transforming monospecific plantations into mixed plantations. Regional examples show how rapidly the promotion of broadleaves under aid schemes for afforestation lost its importance compared to public incentives for the conversion of pure conifer stands (Pieper 2002). The severe storms in the early 1990s in particular fostered the changes in forest policy (which began as early as the 1980s) and, as a consequence, the reforestation practices in both public and private forests. The number of public forest conversion programmes and the related expenditures increased significantly in those years (Schraml and Volz 2004).

Even in the conifer-dominated regions in the northern parts of Europe the amount of broadleaves has increased since the beginning of the 20th century. However, compared with Finland and Sweden, only in Norway has the volume percentage of broadleaved trees increased significantly in the last decades. Analysts find it difficult to relate these tendencies to forest policy measures as Sweden has a long-standing tradition of protecting its temperate broadleaved forests whereas Norway has offered strong silvicultural incentives to promote conifers (Framstad 1996).

Environmental policy

Some valuable broadleaves are characteristic elements of those native forest types that have lost a large percentage of their former cover in Europe. Riverine forests,

Table 5.3. Forestry conditions and changes under Regulation 2080/92 in eight European countries (Institute for Forestry Development 2001).

	DE	DK	ES	FR	IE	IT	PT	UK
				Forestry conditions				
Area under forest (million ha)	10.7	0.4	10.8	13.4	0.6	7.3	3.4	2.6
Forest cover (%)	30.1	12	21.6	24.3	9.2	24.0	36.3	10.7
			Proportion of additional forest areas by type of planting					
All forest types (%)	0.25	0.66	3.80	0.21	19.21	1.21	5.11	5.47
Conifers (%)	0.04	0.01	2.37	0.21	20.92	0.35	3.86	7.29
Broadleaves (%)	0.56	1.60	7.79	0.14	12.80	1.66	5.72	11.40

which are of special concern to environmentalists and public bodies in charge of flood protection, may serve as an example. There is also a call for increased protection of some other vegetation types that includes ash, maple and lime trees (UNEP 2000). The "Habitats" Directive established in 1992 is the main European Community instrument safeguarding biodiversity in Europe. It introduced the obligation to preserve habitats and species of Community interest and, together with the 1979 "Birds" Directive, is designed to help establish a European network of protected sites. Approximately 200 habitat types are classified by the Directive as being of Community interest. Several of the habitats listed in the annex of the Directive include valuable broadleaves. Examples from temperate Europe include Tilio-Acerion forests, thermophilous *Fraxinus angustifolia* woods, alluvial forests with alder and ash and riparian mixed forests containing ash and elm (ETC 2001, EC 1999, 1992).

Moreover, in some of the member nations forest types characterised by valuable broadleaves were protected under national law even before the implementation of the "Habitats" Directive. In Germany, for example, the federal government and the states extended general protection to some forest types typically dominated by valuable broadleaves such as alluvial and riparian forests and specialised montane forests on slopes, screes and in ravines (Winkel and Volz 2003).

Public awareness of forest policy related to valuable broadleaves

In Europe a goal-oriented policy to promote valuable broadleaves is currently restricted to isolated cases and is only found on a national or regional level. Examples include the national promotional guidelines in accordance with Regulation 2080/92 in the Netherlands, the 'Bosdecreet' in Flanders, new silvicultural guidelines in Ireland (Joyce et al. 1998) and several national Nature Conservation laws. In general, the same observation that applies to forest policy as a whole holds true: While there are relatively few regulations aimed directly and exclusively at forestry operations and forests, forestry is widely affected by policy regulations nominally related to the areas of nature conservation, agriculture and economics.

This is yet another reason why the discussion of tree species and forest types usually takes place on the regional level. Because of the changes in landscape caused by afforestation, the protection of certain vegetation types and many aspects of urban forestry, the benefits of valuable broadleaves are best discussed in a local context.

Moreover, even such internationally covered topics as the destruction of tropical rainforests, loss of species diversity and the pollution of forests through emissions can lead to widely different reactions on a regional level. A famous example is the "Waldsterben" (forest die-back) in Germany which in the 1980s became a metaphor for the decline of the environment (Lehmann 1999). In this context the question of tree species becomes highly relevant for two reasons. For many years the scientific literature has pointed out that coniferous trees show a higher susceptibility to smoke. Since the 19th century there have been lists ranking various tree species according to their smoke susceptibility (Heß 1878). Among the most resistant species in those lists are maple (*Acer platanoides, A. campestre*) and alder (*Alnus glutinosa*). Moreover, the "Waldsterben" phenomenon was communicated by the media mainly through pictures of dying fir and spruce trees. While those powerful pictures on the pages of magazines and on television helped to deeply entrench in the public consciousness the message that "conifers are sick", they are expected to do little to foster a special attention toward valuable broadleaves. On the other hand, a somewhat different scenario may result from the public discussion of the destruction of tropical rainforests. The authors expect that pictures of clear-cut tropical forests, which consist of broadleaf trees obvious even to lay observers, may lead to an increased sensibility toward temperate deciduous forests as well (Schraml and Volz 2004).

In summary, we have reached the conclusion that the perception of the significance of valuable broadleaves is mainly formed through the agenda setting of the media and the personal experience value of individual species in the context of forest recreation. Moreover, the possibly different perceptions of different social groups will also have to be considered when examining this process.

5.3.5 Perceptions and preferences

Preferences of experts

As pointed out above, not all people in Europe can be expected to show similar forest preferences. It has been demonstrated repeatedly how much these ideas vary between different social groups and over time (Braun 2000, Lehmann 1999, Riehl 1956). Chapters 2.1 and 5.2 in this book describe this heterogeneity in the areas of nature conservation, economy and forestry; the present chapter is restricted to forest recreation.

The long history of forest aesthetics in Central Europe has generated many guidelines for the assessment and management of forests. The significance of individual tree species to an observer has been described in detail and ranked since recreational suitability

- close to residential
- forest aestethics
  - tree species
  - age/dimension of trees
  - deadwood
- accessibility
  - quality of roads
  - density of roads
- infrastucture
  - other attractions

Quality of Forest Recreation Areas

- Climate
- Topography
- Soils
- Water
- General environment

Figure 5.31. The expert view: Variables used for determining the suitability of forest sites for recreation in Central Europe (Grüning and Kissling-Näf 2000; left) and site factors influencing quality of forest recreation areas in the USA (Douglass 2000; right).

the 19th century. Salisch (1911) declared oak to be the most beautiful tree followed by beech, hornbeam and ash. He goes on to list maple, elm and wild fruit trees before ranking the conifers. Other authors have ranked the various species in a different order, thus demonstrating an unevenness which carries over into the evaluation of broadleaved versus coniferous trees (e.g. Hufnagl 1939).

Aside from the aesthetic arguments that have persisted since the middle of the 19th century, several politically motivated papers have postulated the loss of amenity functions as a consequence of the increase of coniferous forests all over Europe (e.g. Riehl 1855). Non-governmental organisations like the "Committee for saving broad-leaved forests", founded in 1941, developed ambitious efforts to herald the recreational, ecological and cultural value of non-conifers (Münker 1958).

Due to the growing awareness toward recreation since the late 1960s, several approaches to measuring the suitability of forests for recreational purposes have been suggested and applied (Schulz 1978, Gundermann 1972, Ruppert 1971, Scamoni and Hofmann 1969). Since the early days the proportion of tree species in a forest has been a regularly measured indicator even though the "measurement-mania" of the 1970s has since calmed down. From the onset scientists, recreational experts and national heritage organisations widely agreed that broadleaf forests are evaluated more positively than coniferous forests by forest visitors. Several surveys were conducted to demonstrate the congruence between scientific expertise and public preferences. The results of these studies were inconsistent and have been variously interpreted by different authors, except for one fact: in those early days only a small minority (approximately 10%) of the people questioned in Austria, Germany and Switzerland indicated a preference for broadleaved forests (Jacob 1971, Ruppert 1971).

Nowadays the management of forests for recreational purposes is still guided by many suggestions. Nevertheless it remains difficult to identify the role of specific tree species or forest types with respect to the suitability of forests for recreational purposes. Figure 1 gives a current overview of important variables as seen by two experts. Tree species are mentioned in only one of the models, and only as one element among several others.

The experts offer varying opinions as to which factors are considered important to the quality of forest recreation. The emphasis on certain key aspects in describing the ideal recreational forest is influenced by the personal background of the authors. Moreover, there are indications of regional differences. Among forest managers there is a predominance of tree-centred opinions. It is widely believed that in recent years the recreational and ecological discourse has met and found common goals. Furthermore, close-to-nature forestry is considered optimal for forest recreation. Silvicultural experts postulate that "a forest that is managed close to nature widely fulfils the recreational demands of the people" (Leibundgut 1993).

The choice of tree species, silvicultural treatments and regeneration are therefore thought to be the main contributing factors of forest management with regard to recreation (Jacsman 1998). Forest managers are assisted by some stakeholders who emphasise that a large proportion of forested land is suitable for recreation. Depending on the local rules governing forest access, up to 90% of the forests in Central Europe are characterised by a multiple-use forest management that includes recreation (Council of Europe 1995).

There remains a higher level of scepticism among scientists dealing with public preferences in countries with a young forestry tradition such as the U.K. or Ireland. The findings show that a "clear difference between landscapes used for recreation and for timber production" exists and that "the public believes farming and forestry help to maintain the beauty of the environment and acknowledges the need for economic woodlands but prefers such activities either to be located in remote areas or to be softened somehow" (Font and Tribe 2000).

Public perception of forests

The current efforts of the European forest policy to expand the overall forest area and increase forest diversity are hardly reflected in the public perception. The populations of the major European countries view the overall state of Europe's forests as negative. More than 50% of the people in Germany, France, Italy and the United Kingdom are convinced that forest area and species diversity are decreasing and that the health of the forest is in decline. However, the reasons for this unsatisfactory situation are not so much due to the result of activities from within the forestry sector as from external factors, i.e. industrial pollution, traffic and construction activities. Industrial forestry is considered only as a fourth factor responsible for the negative aspects of the situation (Rametsteiner 1999).

The comparatively positive assessment of European forestry in many supra-regional surveys (e.g. Lindholm 2000) does not mean, however, that forest visitors will accept any form of forest management. O'Leary and McCormack (1995) summarise the main results of public opinion polls: (1) fire damaged stands are

Table 5.4. Relevance of species to the evaluation of the forest cover in several European regions. Percentage of interviewees in 16 case studies in 8 countries who indicated that a specific type of forest is relevant for their evaluation of forest cover (results from the MultiforRD project, n = 6471).

Country	AU	DK	DE	GR	HU	IE	NL	ES
Case study area A	15.5	36.3	11.4	13.0	39.6	39.5	19.1	17.2
Case study area B	15.7	27.3	13.0	21.4	32.6	48.3	22.3	25.9

unappealing, (2) clear-cutting is mostly regarded as negative, (3) mature old growth trees are favoured over young forests, (4) an open canopy is preferred, (5) forest openings are favoured and (6) there are no clear preferences for either natural or managed forests. The authors point out that the last item – the lack of a clear preference for either managed or natural forests – gives forest managers the option to plant either broadleaves or conifers. In practice, however, this assessment is subject to strong regional differences.

As part of the Fair funded project MultiforRD in 2000, inhabitants and landowners in 16 European regions were interviewed, among other topics, on the subject of forest perception. The survey first asked for an evaluation of the forest cover in the person's home region. Subsequently, respondents were asked to indicate if their answers to this question were dependent upon a special type of forest. Table 5.4 shows the significant differences between the answers of participants in several European regions.

Overall, Central Europeans were less concerned with the issue of species, while inhabitants of countries with a high level of recent afforestation showed a much greater sensitivity toward this subject. Similar results could also be obtained on a regional level within different countries. In some countries there are remarkably high regional differences. While the results in Germany, Austria and the Netherlands are relatively homogenous, case studies in Denmark, Greece, Spain and Ireland reveal noticeable regional differences within the countries.

In Ireland, people who live in traditional forestry areas show a more positive disposition toward coniferous forests than those in areas with lower forest cover and a high level of afforestation activity. Feelings of resentment toward conifers are correlated with a lower level of knowledge regarding the differences between broadleaved and coniferous trees in the less forested areas. In those areas, between one fifth and one quarter of the respondents were unable to classify certain tree species (oak, spruce, ash, pine and beech) as either broadleaves or conifers when the species names were mentioned in the interview or when the subjects were shown pictures of the trees (O'Leary et al. 2000).

In regions with a long-standing forestry tradition, the results of methodically similar surveys differ greatly. Here a clear majority of the interviewees will recognise at least the main tree species of their region. For example, detailed studies from Germany reveal that, among the broadleaf trees, the cherry is recognised by a large number of people while only a minority is able to correctly identify the ash. Overall, the degree of recognition, and thus also the value assessment, of valuable broadleaves ranks clearly below that of oak, beech and birch (Rozsnyay 1979).

As mentioned above, numerous surveys dealing with public preferences of forest types are available. There are several compilations of such surveys for the German-speaking countries (Schraml 1999, Rozsnyay 1972, Jacob 1971, Ruppert 1971; for France compare Deuffic and Lewis 2004). While the existing time sequence must be interpreted with caution since unequal approaches and methods were used, it is apparent that the percentages of the four frequently used categories for testing preferences ("broadleaf forests", "conifer forests", "mixed forests" and "no opinion") change over time. The popularity of mixed forests remains at a steady level whereas conifer forests have decreased significantly in popularity to the benefit of broadleaf forests. While opinion polls in Germany in the 1970s regularly indicated a preference for conifers among approximately 25% of the poll participants (Rozsnyay 1972), this number has decreased to 15% in the latest nation-wide survey (Schraml and Volz 2004). At the same time a remarkable number of people have no opinion on this issue, i.e. the subject is of no interest to them. In most of the German studies of the 1960s and 1970s in which the category "no opinion" was offered, between 20% and 50% of the interviewees chose this answer. Today the number of undecided people appears to be smaller.

In this regard the results are similar all over Europe. In a study in the U.K. people were asked about their tree-type preferences in local forests. 37% of the respondents liked any type of tree, 33% preferred the mixture of conifers and broadleaves where broadleaves were more often named as favourite (22%) than conifers (5%) (Lee 2001).

This high value assessment of mixed forests may serve as an argument for the importance of the role of valuable broadleaves as frequently planted admixed species. Whether people are truly aware of this role, however, must be evaluated under consideration of regional and social differences against the background of the quoted empirical results with regard to species recognition.

Attempts to characterise the group of people that show a clear preference for broadleaf trees demonstrate repeatedly that age, education and origin (rural versus urban areas) are factors of high relevance with regard to existing preferences (Lee 2001, Braun 2000, Wöbse 1972). Overall it can be said that members of the well-educated classes with a higher social status tend to prefer broadleaf trees. This appears to hold true for urban areas as well. There are also indications that people who frequently visit the forest tend to prefer broadleaf trees. Finally, as can be expected, members of environmental organisations also show a marked preference for broadleaves. However, since several of the above variables correlate with certain social groups, these results are not surprising (Table 5.5).

There is another interesting factor that seems to explain the preferences of many citizens. People in regions where broadleaves are already numerous tend to show a preference for broadleaf trees. This correlation, which has been described on the basis of individual examples from various regions in Europe (Vanderlinden and Lust 1998, Ott 1980, Degener 1963), was most recently examined systematically in Germany (Schraml and Volz 2004). A comparison of the preferences of people from states with a high percentage of broadleaves with those from regions with fewer

Table 5.5. Variables perceived to have an effect on the preference for broadleaves.

Socio-demographic variables	Personal forest use
Age Formal Education Social class Urban surrounding Membership in environmental organisation	Use of wood and other tree parts As a forest owner Recreational use Number of forest visits
Forest characteristics	**Forest policy**
Proportion of broadleaves in home region Proportion of broadleaves on own property	Significant landscape changes (i.e. afforestation)

broadleaves reveals obvious differences. People from broadleaf-rich regions show a higher preference for broadleaves than those from broadleaf-poor areas. It is interesting to note that another survey question that rates the "satisfaction with the state of the forest in the region" is not affected by the proportion of broadleaves in that region (see also Lafitte 1993).

The adaptation of preferences to local conditions is a phenomenon that could also be confirmed for small forest owners. There is a positive correlation between the given tree species composition in forest holdings and the preferences of the owner (Bieling and Schraml 2004). In general, forest owners play a special role compared to other social groups. A comparison of the results of the MultiforRD study from several European countries reveals that forest owners clearly demonstrate a higher sensitivity to the significance of forest types with regard to forest quality than is shown by other groups. Similar results can be found in several national studies. Forest owners in Germany, for example, more frequently profess a preference for coniferous forests than do other people (Schraml and Volz 2004).

Unlike the afforestation debate, in which species aspects play an important role, the topic of forest conversion has never caused much public resonance. In Germany 33% of the population has never heard about the efforts to increase the percentage of broadleaf trees in the forest (Schraml and Volz 2004). Surveys conducted in several European countries show that most people do not notice this change even if questioned in the forest or else suspect other reasons for the changes in the forest (Kobel 1997, Meierjürgen 1994, Oesten and Roeder 1994). The "forestry revolution" toward a close-to-nature forestry occurs largely unnoticed by the public (Schriewer 1998).

Connotations of forest types

Different associations connected with forest types are described all over Europe. Table 5.6 summarises associations found in a German study. Lee (2001) reports from Scotland that coniferous forests are associated with "hillsides really closed in by forests." People complained that conifers do not permit easy access for walking whereas broadleaf forests are forests "where you can just wander and you can make

Table 5.6. Perception of coniferous and broadleaf forests in Germany (n=50 deep side interviews; Schriewer 1998).

Conifer Forests	Broadleaved Forests
artificial, man-made	native, natural
darkness, black	light, diverse colours
impermeable, repelling	permeable, inviting
uniform, military	individual
monotony	diversity
young trees	old trees
mushrooms	flowers
---	cathedrals
"waldsterben"	---

your own way." Typical results are also shown by several Irish surveys. The interview participants named the following as principal reasons for their preference for broadleaf trees: "It looks the best" (54%) and "not enough broadleaves are being planted at present" (12%). Only 7.5% of the respondents explicitly mentioned wildlife or species diversity as an explanation for their preference for broadleaves (O'Leary et al. 2000).

Most of the survey participants prefer broadleaf forests for aesthetic reasons and because of their supposedly easier accessibility. Ecological reasons for broadleaf preference are rarely mentioned by laymen. In many cases the preference for broadleaves simply results from a rejection of coniferous trees without a clear statement as to the advantages of broadleaves.

An explanation for this rejection of coniferous forests and the relative preference of deciduous and mixed forests can also be found in the connotation of terms relevant to forest policy. This is shown in a representative German study which asked for the associations between various terms and certain forest types (Figure 2).

Classical forest functions associated with terms such as "naturalness" and "habitat for plants and animals" are mainly connected with the category "mixed forest." While "deciduous forest" is typically associated with the topics of forest recreation and nature conservation by about one third of the interviewees, it, unlike the other two forest types, does not have a profile dominated by particular contents. This suggests that a classical broadleaf forest motif is not part of the public consciousness. Instead, the high level of preference for deciduous forests is mainly based on the rejection of coniferous forests.

Approximately 40% of the surveyed individuals made a connection between forestry and coniferous forests. Thus "forestry", together with "monoculture" and "waldsterben", belongs to those categories least associated with broadleaf forests. Many forest visitors, their sensibilities heightened by media reports, are able to notice the presence of diseased conifers while failing to recognise damage to the "strong and green" broadleaf trees (Schriewer 1998).

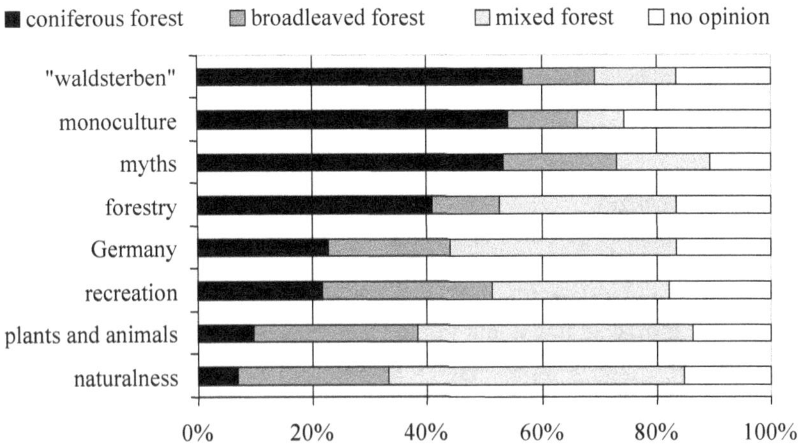

Figure 5.32. Associations of various terms with forest types. Representative sample in Germany (n=2822; Schraml and Volz 2004).

5.3.6 Conclusions

Main findings

The public perception of valuable broadleaves presents itself as highly differentiated. On the one hand this group of tree species profits from the trend in European forestry toward increased diversity. The implementation of international agreements and conventions has met with ongoing efforts to foster the amenity functions of European forests. While there are large differences between the national and regional approaches to this problem, as well as within the individual policy programmes related to valuable broadleaves, their overall direction is unmistakeable. The economic, ecological and social roles of valuable broadleaves are recognised and the planting of these species is supported through financial incentives while clear-cutting of forests containing broadleaves is often prohibited by law.

At the same time the lack of a nominal "valuable broadleaves policy" becomes apparent. A direct promotion of these tree species is only found in a few regional cases. The various concepts, regulations, laws and aid schemes in relation to valuable broadleaf forests fall into such different categories as agriculture (afforestation), environment, close-to-nature forestry, flood prevention and recreational aspects of urban forestry. There is no co-ordination of these policies with regard to the planting of valuable broadleaves. On the contrary, competing laws and regulations often impede the extension of the broadleaf forest area. The example of the implementation of the EU afforestation regulation illustrates particularly well the varying extent to which different regions take advantage of the potential of valuable broadleaves and the policies that promote their planting.

Differences are also found in the value assessment of valuable broadleaves by the citizens. While there are indications of a common trend, regional peculiarities are quite apparent. There are no widespread preferences for pure broadleaf forests among the people of Europe, although the rejection of pure coniferous forests for aesthetic and ecological reasons is now so widespread in Europe that a mixture of different tree species is clearly preferred. Any temporal series of survey results will illustrate these changes.

Finally, the presented studies on people's forest preferences verify the theories introduced and the assumptions made in explaining these preferences. The arguments used to explain the formation of preferences lend plausibility to the preference for a diverse forest and to the changes in the evaluation of coniferous forests based on available information. It could be shown that it is mainly the rising awareness of ecological problems in general that helped broadleaf trees to become popular. It is also confirmed that personal preferences are indeed influenced by individual and group-specific experiences. Unlike the general state of the forest, the personally known forests near a person's home are often evaluated very positively regardless of their tree species composition. The adaptation of preferences is a phenomenon illustrated by many studies all over Europe.

Today the support for valuable broadleaves is mainly based on the arguments of experts for a specific species composition of forests. For the general public, species are a less relevant factor used to describe forest quality. Arguments drawn from policy discussions and traditional myths surrounding several species rarely reach the general public. Both are restricted to a group of insiders.

Moreover, one phenomenon shows that this situation is not static. Whereas people tend to adapt their preferences to a given state of the forests, the sensitivity to species was successfully stimulated in Europe's afforestation areas. In the regions where particularly extensive landscape changes due to afforestation have occurred, the resistance to afforestation is mainly influenced by the question of tree species selection. The coincidence of environmental debates and aesthetic concerns made tree species a crucial issue of local policy and encouraged the people's demand for broadleaf trees.

Towards a valuable broadleaves policy

The significance of lay opinions in forestry has repeatedly been a subject of discussion. Already the question of preferences for certain tree species and forest types in surveys has been criticised (Heeg 1971). An inherent problem of these studies is the realisation that a question may be considered controversial by experts but not by lay persons. The issue of tree species, while crucial to the experts, is only of very limited interest to the general public. In many other cases it is questionable whether the interviewees actually have a clear concept of the forest types on which they are asked to comment.

The actual differences between the assessments by forest experts and local forest users have been demonstrated repeatedly (Coles and Bussey 2000). In regard to the criteria used for describing the value of valuable broadleaf forests, this difference is illustrated by the fact that experts describe forests in highly species-specific terms while laymen do not. This is also reflected in the language where universally accepted technical terms on the supra-regional level stand in contrast with the often locally restricted terms and individual codes of laymen.

Indeed the question "Do species matter?" may not be answered uniformly by all forest visitors. This is also supported by a summary of the perception of forests as identified in different research reports in Central Europe (Schmithüsen and Wild-Eck 2000). The different perceptions of forests described in this summary range from views in which the question of tree species presumably plays a minor role (forests as places for movement and activities, for breathing air, for health, for retreat), to others that may be strongly influenced by the existing tree species (forests as a symbol of origin, mystical space, sensual space and a spiritual place).

In order to adequately describe the observed preferences of people, some authors revert to a distinction between "manifest preferences" and "true preferences." Manifest preferences are actual preferences, as shown in recreational behaviour with respect to forest types or beliefs that are revealed in interviews. True preferences are those a person would develop if he or she had all the relevant and available information. This information includes the benefits of different forest types and tree species in addition to the experience and aesthetic value of these forests. Unfortunately this distinction between informed and uninformed persons does not give credit to the fact that a significant percentage of the so-called laymen have made their own experiences with the forest.

While these lay persons may be lacking in concentrated knowledge on the issue of tree species, some of them possess qualities often not equally shared by the experts, i.e. the knowledge of the suitability of certain forest types for sensual, spiritual, mystical or symbolic purposes. This qualitative knowledge in turn makes them experts on each of these topics. Whether the views of these lay persons can be incorporated into the forest management ultimately remains the decision of politicians, interest groups and forest owners. Nevertheless, the consideration of the public's wishes leads to communication with the citizens through which laymen and experts can learn from each other. Due to their economic, ecological and social attraction, valuable broadleaves are particularly well suited to foster discussion between these various groups. Therefore valuable broadleaves have the chance to step out of the shadow of conifers. They are already regarded as being more valuable than conifers in an environmental and aesthetic sense. In addition to this positive evaluation, valuable broadleaves should be given their own positive and authentic connotations. Modern myths surrounding these trees can be found in the fields of recreation, ecology and heritage. Forest scientists can help to describe, disseminate and, as a consequence, revitalize them.

In this context the German folklorist Lehmann (1999) suggested an interesting hypothesis concerning an increasing enthusiasm for broadleaves in general and valuable broadleaves in particular. Since a society's view towards forests changes with the development of society, and since there is a current societal trend toward individualisation, Lehmann expects an individualistic view of forests. He states that "the process of individualisation appears to go hand in hand with a devaluation of the symbol "forest" ". Enlightened individuals like to find themselves reflected in other individuals, namely the "trees". The special suitability of valuable broadleaves for personal identification has already been mentioned; they are currently undergoing the process of becoming "culturally privileged."

References

Appleton, J. 1975 The experience of Landscape. John Wiley and Sons, London.

Balling, J.D. and Falk, J.H. 1982 Development of visual preference for natural environments. Environment and Behaviour 1: 5–28.

Bieling, C., Schraml, U. 2004 Was ist der Natur näher als der Wald. Zur Wahrnehmung des Waldzustandes durch private Eigentümer. Allgemeine Forst- und Jagdzeitung 3: 41–48.

Braun, A. 2000 Wahrnehmung von Wald und Natur. Leske und Budrich, Opladen.

Buchanan, T., Christensen, J.E. and Burdge, R.J. 1981 Social Groups and the Meaning of Outdoor Recreation Activities. Journal of Leisure Research 13: 254–266.

Coles, R.W. and Bussy, S.C. 2000 Urban forest landscapes in the UK – progressing the social agenda. Landscape and Urban Planning 52: 181–188.

COM, Commission of the European Union 1998 Communication from the Commission to the Council and the European Parliament on a forestry strategy for the European Union. COM (1998) 649, 03/11/1998. Http://europa.eu.int/comm/agriculture/fore/comm/649_en.pdf. (last visited 21.06.2002)

Council of Europe 1995 Forests in Europe: Proceedings from the 4th Pan-European Colloquy on Tourism and the Environment, Warsaw, 20–21 September 1994. Council of Europe Press, Strasburg.

Degener, P. 1963 En Undersogelse af Publikums Krav til Skoven. Dansk Sovforenings Tidsskrift 48: 489–548.

Deuffic, P. and Lewis, N. Rural forests under the influence of urban values. Sociological analysis of the link "forest-socity" in France (1960–2005) Paper presented to the XI World Congress of Rural Sociology, July, 25-30 2004, Trondheim, Norway.

Deutsches Baumarchiv 2004 Http://deutschesbaumarchiv.de (last visited 20.09.2004)

Dewey, J. 1958 Art as Experience. Capricorn, New York.

Douglass, R. W. 2000 Forest recreation. Waveland Press, Prospect Heights.

EC, European Commission, DG Environment 1992, Council Directive 92/43/EEC of 21 May 1992 on the conservation of natural habitats and of wild fauna and flora, Annex 1 Http://europa.eu.int/comm/environment/nature/hab-an1en.htm. (last visited 21.06.2002)

EC, European Commission, DG Environment 1999 Interpretation Manual of European Union Habitats EUR 15/2, October 1999. Http://europa.eu.int/comm/environment/nature/habit-en.pdf. (last visited 21.06.2002)

Elster, J. 1983 Sour Grapes: Studies in the Subversion of Rationality. Cambridge University Press, Cambridge.

ETC, European Topic Centre on Nature Conservation, European Environment Agency 2001 Alpine region: Reference list of habitat types and species present in the region. Http://nature.eionet.eu.int/activities/Natura_2000/Ref-alp (last visited 21.06.2002)

Font, X. and Tribe, J. (eds.) 2000 Forest tourism and recreation: case studies in environmental management. CABI Publ., Wallingford.

Framstad, K.F. 1996 Environmental effects of public forestry incentives in Finland, Norway and Sweden. Journal of Forest Economics 2/3: 289–313.

Freiburger Anthologie 2002, Http://www.freiburger-anthologie.de, (last visited 11.06.2002)

Großherzogliches Ministerium der Finanzen, Abteilung für Forst- und Cameralverwaltung, ed, 1904, Bemerkenswerte Bäume im Großherzogtum Hessen in Wort und Bild. Zedler und Vogel, Darmstadt.

Grüning, C. and Kissling-Näf, I. 2000 Die IFRI Forschungsstrategie und ihre Anwendbarkeit für eine Untersuchung in der Schweiz. Professur für Forstpolitik und Förstökonomie, Department Holzwissenschaften, ETH Zürich, Arbeitsberichte 00/2, Zürich.

Gundermann, E. 1972 Untersuchungen zur Erfassung, Wertung und Ordnung der Erholungsfunktion von Waldbeständen im bayerischen Hochgebirge. Forstliche Forschungsanstalt München, Forschungsberichte 4, Frank, München.

Harrison, R.P. 1992 Forests. The Shadow of Civilization. University of Chicago Press, Chicago.

Heeg, B. 1971 Meinungsbefragungen im Bereich der Forstpolitik. Forstarchiv 11: 217–223.

Hellström, E. 1996 Environmental forestry conflicts, forest policies and the use of forest resources: recent developments in USA, Germany, France, Sweden, Finland and Norway. EFI Working paper 7. European Forest Institute.

Hellström, E.and Welp, M. 1996 Environmental forest conflicts in Germany – from national to international concern. EFI Working paper 11. European Forest Institute.

Hellström, E. and Rytilä, T. 1998 Environmental forest conflicts in France and Sweden: Struggling between local and international pressure. EFI Working paper 16. European Forest Institute.

Heß, R. 1878 Der Forstschutz, Teubner, Leipzig.

Hilger, H.1956 Geheimnis des Baumes. Herder, Freiburg.

Hook, S. 2002 Impacts of Perception on the paradigm of Sustainable Development. Paper presented at the Meeting of Forest Policy Scientists in 20.–22.03.2002, Sopron, Hungary.

Hufnagl, L. 1939 Die Waldschönheit und ihre Pflege. Springer, Wien.

Institute for Forestry Development 2001 Evaluation of the Community aid scheme for forestry measures in agriculture of Regulation No 2080/92, Http://europa.eu.int/comm/agriculture/eval/reports/forest/text_en.pdf (last visited 21.06.2002)

Jacob, H. 1971 Bemerkungen zu waldbaulich-ästhetischen Problemen in Erholungswäldern. Natur und Landschaft 7: 181.

Jacsman, J. 1998 Konsequenzen der intensiven Erholungsnutzung für die Wälder im städtischen Raum. Schweiz. Zeitschrift für Forstwesen. 6:423–439.

Joyce, P.M., J. Huss, R. McCarthy, A. Pfeifer and E. Hendrick 1998 Growing broadleaves – silvicultural guidelines for ash, sycamore, wild cherry and oak in Ireland. COFORD, Dublin.

Kaplan, S. 1987 Aethetics, affect, and cognition: Environmental preference from an evolutionary perspective, Environment and Behavior 1: 3–32.

Kaplan, S. 1992 Environmental Preference in a Knowledge-Seeking, Knowledge-Using Organism, In: Barkow, J.H.; Cosmides, L. and Tooby, J. (eds.). The Adapted Mind, Evolutionary Psychology and the Generation of Culture. Oxford University Press, New York. Pp. 581–598.

Kearney, A., Bradley, G., Kaplan, R. and Kaplan, S. 1999 Stakeholder perspectives on appropriate forest management in the Pacific Northwest. Forest Science 1: 62–73.

Keith, T. 1983 Man and the Natural World. Changing Attitudes in England 1500–1800. Penguin, Harmondsworth.

Kennedy, J., Thomas, J. and Glueck, P. 2001 Evolving forestry and rural development beliefs at midpoint and close of the 20th century. Forest Policy and Economics 3: 81–95.

Klein, L. 1908 Bemerkenswerte Bäume im Großherzogtum Baden (Forstbotanisches Merkbuch) Winter, Heidelberg.

Kobel, C. 1997 Forstliche Maßnahmen in der öffentlichen Wahrnehmung am Beispiel des Hardwaldes in Olten. Schweiz. Zeitschrift für Forstwesen 7: 533–540.

Konijnendijk, C. 1999 Urban forestry: overview and analysis of European forest policies – Part 2. Europen Forest Institute, EFI Working paper 20, Joensuu.

Konijnendijk, C.C., Nilsson, K., Randrup, T.B. and Schipperijn, J. (Eds.) 2005 Urban Forests and Trees. A Reference Book. Springer, Berlin.

Kühn, S., Ullrich, B., Kühn, U. 2002 Deutschlands alte Bäume. BLV Verlag, München.

Lafitte, J.-J. 1993 Sondage d'opinion sur les forets périurbaines. For. Fr. 4: 483–493.

Lee, T. 2001 Perceptions, Attitudes and Preferences in Forests and Woodlands. Forestry Commission, Edinburgh.

Lehmann, A. 1999 Von Menschen und Bäumen: Die Deutschen und ihr Wald. Rowohlt, Reinbek bei Hamburg.

Leibundgut, H. 1993 Der Erholungswald. Schweiz. Zeitschrift für Forstwesen 1: 41–48.

Lindholm, A. 2000 Nordic Forstry. The public opinion in Holland, Germany and Great Britain. Http://www.smy.fi/tiedotteet/Demoskop99.pdf. (last visited 21.06.2002)

Marzell, H. 1929 Die kultische Verehrung der Bäume. Naturschutz 7: 193–195.

Marsden, T., Milbourne, P., Kitchen, L., Bishop, K. 2003 Communities in Nature: The Construction and Understanding of Forest Natures. Sociologica Ruralis 3: 238–256.

Mayer-Gampe, P. 1999 Der Wald als Symbol in Märchen und Mythen. Dissertation. Universität München.

MCPFE, Ministerial Conference on the Protection of Forests in Europe 1998 Third Ministerial Conference on Protection of Forests in Europe, 2–-4 June 1998, Lisbon, Portugal. Http://www.minconf-forests.net/Basic/FS-MCPFE-Resolution.html. (last visited 21.06.2002)

MCPFE, Ministerial Conference on the Protection of Forests in Europe 2001 The MCPFE and the IPF/IFF Proposals of Action. MCPFE Paper 1, May 2001. Liaison Unit Vienna, Austria. Http://www.minconf-forests.net/Pdf/mcpfe_paper1_ipf-iff.pdf. (last visited 21.06.2002)

MCPFE, Ministerial Conference on the Protection of Forests in Europe 2003 Fourth Ministerial Conference on Protection of Forests in Europe, 28–30. April 2003, Vienna, Austria. Declaration and Resolutions. Http://www.mcpfe.org (last visited 25.09.2004)

Meierjürgen, U. 1994 Freiraumerholung in Berlin: die Entwicklung eines forstpolitischen Ansatzes für eine Raumplanung des stadtnahen Erholungswaldes, Dissertation. Universität Freiburg.

Miller, R. W. 1988 Urban forestry: planning and managing urban greenspaces. Englewood Cliffs, NJ, Prentice- Hall.

Münker, W. 1958 Dem Mischwald gehört die Zukunft. Ausschuß zur Rettung des Laubwaldes im Deutschen Heimatbund. Deutscher Heimatverlag, Bielefeld.

Oesten, G. and Roeder, A., eds. 1994 Zur Wertschätzung der Infrastrukturleistungen des Pfälzerwaldes. Mitteilungen aus der FVA Rheinland-Pfalz 27/94, Trippstadt.

O'Leary, T. and McCormack, A. 1995 Public perception of forestry – where to next? In: Hyttinen, P., Kähkönen, A. and Pelli, P. (eds.). Multiple use and environmental values in forest planning: Proceedings of the International Summer Course, 5–10 June 1995, Tohmajärvi, Finland. EFI Proceedings 4. European Forest Institute. Pp. 133–146.

O'Leary, T., McCormack, A. and Clinch, J. 2000 Afforestation in Ireland – regional differences in attitude. Land Use Policy 17: 39–48.

Orians, G.H. 1986 An ecological and evolutionary approach to landscape aesthetics. In: Penning-Rowsell and E.C., Lowenthal, D (eds.). Landscape meanings and values. Allen & Unwinn, London. Pp. 3–25.

Orians, G.H. and Heerwagen, J.H. 1992 Evolved Responses to Landscapes, In: Barkow, J.H., Cosmides, L. and Tooby, J. (eds.). The Adapted Mind, Evolutionary Psychology and the Generation of Culture. Oxford University Press, New York. Pp. 555–579.

Ott, W. 1980 Wald und Forstwirtschaft im Spiegel der öffentlichen Meinung. Allgemeine Forstzeitschrift 15: 397–399.

Philpot J.H. 1897 The sacred tree or the tree in religion and myth. Macmillan, London.

Pieper, S. 2002 Durch Förderung gesteuerte waldbauliche Entwicklung des Privatwaldes im Bereich der Landwirtschaftskammer Hannover. Forst und Holz 10: 309–312.

Piussi, P. and Pettenella, D. 2000 Spontaneous Afforestation of Fallows in Italy. In Weber, N (ed.). NEWFOR – New Forests for Europe: Afforstation at the Turn of the Century. Proceedings of the Scientific Symposium February 16-17, Freiburg, Germany. EFI Proceedings 35. Pp. 151–163.
Rametsteiner, E. 1999 The attitude of European consumer towards forests and forestry. Unasylva 50: 42–48.
Ribe, R. 1989 The aesthetics of forestry: what has empirical preference research taught us? Environmental Management 1: 55–74.
Ribe, R. 1999 Regeneration Harvests Versus Clearcuts: Public Views of the Acceptability and Aesthetics of Northwest Forest Plan Harvests. Northwest Science 73: 102–117.
Riehl, W.H. 1855 Die Naturgeschichte des Volkes als Grundlage einer deutschen Sozial-Politik. Cotta, Stuttgart.
Riehl, W. H. 1956 Deutscher Wald. Langewiesche, Königstein im Taunus.
Rozsnyay, Z. 1972 Ergebnisse eines Jahrzehnts forstlicher Meinungsumfragen. Forstarchiv 8: 149–159.
Rozsnyay, Z. 1979 Kennen die Stadtbewohner die Waldbäume? Allgemeine Forstzeitschrift 1/2: 26–28.
Ruppert, K. 1971 Zur Beurteilung der Erholungsfunktion siedlungsnaher Wälder. Mitteilungen der Hessischen Landesforstverwaltung, 8, Sauerländer, Frankfurt/Main,
Salisch, H. v. 1911 Forstästhetik. Julius Springer, Berlin.
Scamoni, A. and Hofmann, G. 1969 Verfahren zur Darstellung des Erholungswertes von Waldgebieten, Archiv für Forstwesen 3: 283.
Schmithüsen, F. and Wild-Eck, S. 2000 Uses and Perceptions of Forests by People Living in Urban Areas: Findings from Selected Empirical Studies. Forstwissenschaftliches Centralblatt 6: 395–408.
Schraml, U. 1999 Meinungsforscher im Wald – 30 Jahre Umfrageforschung als Basis einer aktuellen Untersuchung. In: Zusammenstellung der fachlichen Beiträge präsentiert am Jahrestreffen der deutschsprachigen Lehrstühle für Forstpolitikwissenschaft, 6–9 April. In: Siders, Schweiz. Eds. Zimmermann, W., Volken, T. Professur Forstpolitik und Forstökonomie, ETH Zürich – Grundlagen und Materialien 3, Zürich. Pp. 4–12.
Schraml, U. 2003 Expectations towards forestry. The influence of personal networks with forest owners. Urban Forestry and Urban Greening 3: 161–170.
Schraml, U. and Volz, K.-R. 2004 Conversion of coniferous forests in social and political perspectives. Findings from selected countries with special respect to Germany. In Spiecker, H., Hansen, J., Klimo, E., Skovsgaard, J.P., Sterba, H. and Teuffel, K. v. (eds.). Norway Spruce Conversion – Options and Consequences. EFI Research Report 18, S. Brill, Leiden, Boston, Köln. Pp. 97–119.
Schriewer, K. 1998 Die Wahrnehmung des Waldes im Wandel. Vokus 2: 4–17.
Schulz, H.-J. 1978 Naherholungsgebiete: Grundlagen der Planung und Entwicklung. Parey, Berlin.
Sebillot 1906 in Marzell, H. 1929 Die kultische Verehrung der Bäume. Naturschutz 7: 193–195.
Sommer, R. and Summit, J. 1996 Cross-National Rankings of Tree Shape. Ecological Psychology 4: 327–341.
Sommer, R. 1997 Further Cross-National Studies of Tree Form Preference. Ecological Psychology 2: 153—160.
UNCED, United Nations Conference on Environment and Development 1992 Report of the United Nations Conference on Environment and Development, Rio de Janeiro, 3–14 June 1992, Annex III, Non-legally binding authoritative statement of principles for a global consensus on the management, conservation and sustainable development of all types of forests, Http://www.un.org/documents/ga/conf151/aconf15126-3annex3.htm. (last visited 21.06.2002)

UNEP, World Conservation Monitoring Centre, 2000 European Forests and Protected Areas: Gap Analysis, Cambridge, UK, Http://www.unep-wcmc.org/forest/eu_gap/ Technical%20Report.pdf. (last visited 21.06.2002)

Vanderlinden, I. and Lust, N. 1998 Kenntnis und Einstellung der Bevölkerung in Bezug auf den Wald in relativ waldreichen und waldarmen Regionen in Flandern, Paper presented at the Meeting of Forest Policy Scientists, 01.- 03. 04. 1998, Freising, Germany.

Wild-Eck, S. 2002 Statt Wald – Lebensqualität in der Stadt. Die Bedeutung naturräumlicher Elemente am Beispiel der Stadt Zürich. Seismo, Zürich.

Wilson, E.O. 1984 Biophilia: The Human Bond with Other Species. Harvard University Press, Cambridge, MA.

Winkel, G. and Volz, K-R. 2003 Naturschutz und Forstwirtschaft: Kriterienkatalog zur "Guten fachlichen Praxis". Schriftenreihe „Angewandte Landschaftsökologie", 52, Landwirtschaftsverlag, Münster-Hiltrup.

Wöbse H.-H. 1972 Untersuchung zum Nutzungs- und Bestandswandel der Solling-Wälder. Die vom Urlauber bevorzugten Waldtypen und Folgerungen für die Planung. Diss. TU Hannover.

6. Conclusions

6.1 Basic Features for Growing Valuable Broadleaved Trees in Europe

Heinrich Spiecker
Institute for Forest Growth, University of Freiburg, Germany

Valuable broadleaved species are considered to be valuable because of their high timber prices, and aesthetics of their timber, or simply because of rareness and beauty of the tree. Which species really is valuable, is often a matter of subjective opinion. However, common ash (*Fraxinus excelsior* L.), sycamore (*Acer pseudoplatanus* L.) and wild cherry (*Prunus avium* L.) are considered to be the most important representatives of the group of valuable broadleaved species with still increasing economical importance. Not only these species, but also several other valuable broadleaves such as walnut (*Juglans regia*, *J. nigra* and hybrids), wild service tree (*Sorbus torminalis*), black alder (*Alnus glutinosa*) and lime (*Tilia cordata*) are adapted to a wide range of sites in Europe. They can be grown successfully as part of an existing forest management regime, in private and public forests and in a mixture with both broadleaved and coniferous trees. Furthermore, they can be grown as a result of afforestation of farmland, in orchards, along roads or in hedge rows combined with agricultural land use, either arable land or grazing land. Unregulated human interventions have changed the genetic composition of valuable broadleaved stands. Recent efforts in tree breeding and genetic conservation help to improve the situation.

Ecological and economic considerations recently increased the interest in growing valuable broadleaved tree species. Timber quality depends on the genetic characteristics of the individual tree as well as on management. Applying an approriate treatment they can yield high quality timber within relatively short production times. Valuable broadleaved forests may form important habitats for numerous plants, insects, fungi and animals and so increase habitat values, as well as contribute to biodiversity and nature conservation.Valuable broadleaves produce flowers in springtime, their fruits enrich habitats, and they contribute with additional colour and texture to the beauty of the landscape and increase the attractiveness of forests for recreational use. In natural conditions, valuable broadleaves grow scattered as single trees and in groups. Apropriate management including the formation of internal edges and gaps may change the visual impacts and create high aestethic values.

However, contribution to landscape varies from place to place. Valuable broadleaves mixed with other species in groups or small stands reduce risks of pests and diseases. When planted in huge monocultures risks of pest and diseases may increase, on proper sites, mixed with other tree species or in hedge rows these risks are reduced. There may be a wide range of management aims to be acomplished within the same forest. The aims vary considerable in different regions and they change over time. Diversification and flexible management has to scope with these challenges. Natural processes help to reduce cost and precise management concentrating on the production goals still produces high values.

The public preference as well varies locally and depends on the social background of the people. Even so there is a trend towards more "broadleaved" forest, the public does not emphasize to a high extent valuable broadleaved tree species but clearly prefers mixture of species. Size and structure of the forest is more important than species composition.

6.2 Future Strategies for Growing Valuable Broadleaved Trees in Europe

Heinrich Spiecker
Institute for Forest Growth, University of Freiburg, Germany

General silvicultural strategies are difficult to develop for valuable broadleaves, because they are a heterogeneous group and species reactions differ between management regimes and site conditions. One common feature is that they generally require regular release from competitors on most sites for survival and optimal growth. As many valuable broadleaves have a limited capacity to compete in forests, they require more interventions especially when mixed with fast growing and shade tolerant tree species. When aiming for high wood quality in a relatively short time, productive sites are required. The site conditions modify growth dynamics including height growth and crown architecture while diameter growth in forests is mainly affected by the growing space. The dimension and wood properties of the clear bole determine the value of the crop. These parameters can be controlled by forest management such as initial spacing, thinning, pruning and the time of harvest. Stem form and growth characteristics are as well partly predetermined by genetic properties and quality of the plant material. When planting, only a small number of genetically well selected and site adapted trees are needed. Recent studies on crown architecture help to understand the dynmics of crown development and crown shape as a result of genetic properties, site conditions and treatment. These studies contribute to improving artificial pruning and thinning. The spacing design is influenced by the need of selecting best performing trees out of a larger nunber of trees, the landscape and technical aspects. Geometric forms such as rows facilitate the access to the trees and the use of machinery, on the other hand naturally shaped, diffuse edges in an irregular and asymmetric way may enhance the beauty of a landscape. Valuable broadleaves as well regenerate naturally from seeds or root suckers. Young trees may need to be protected against mice as well as against browsing and bark stripping by red deer, grey squirrel or other animals. Naturally regenerated minority tree species in mixed forests offer an often underestimated potential for growing valuable timber when properly managed.

Most valuable broadleaved tree species are light demanding and grow in forests with varying canopy closure. As competition induced natural pruning can be replaced by artificial pruning of open grown trees, alternative management options offer innovative ways for the production of valuable wood. These management options may at the same time also increase the supply of non-wood products and services. Valuable wood production has to concentrate on the individuals which are expected to produce the high quality wood in the wanted dimension. In order to improve management efficiency interventions have to be limited to actions, which support those value producing individuals.

Future crop trees should be selected, when the future development can be predicted and measures for supporting the development of the future crop trees are needed. Such a measure is artificial pruning. The timing of pruning has an impact on the stem shape and development of the remaining branches. The number of future crop trees per ha depends on the target diameter and the production time: the larger the diameter and the shorter the production time the fewer trees per ha should be selected as future crop trees. Artificial pruning has to be repeated in order to avoid pruning of large branches, to reduce the impact of pruning on the tree and to reduce the size of the knotty core inside the trunk. Selective pruning of the biggest branches should be preferred to (pseudo-) whorl wise pruning.

Moreover, thinning is applied for controlling the quality of the wood production; in particular, it is used for favoring future crop trees, controlling their diameter growth and natural pruning. A two-phase management system is recommended: first phase emphasizing pruning, and second phase encouraging crown expansion and stimulating diameter growth. There is a close relation between crown expansion and diameter growth. An allometric model describing the relation between crown width and stem diameter is used to calculate the intensity of thinning in order to control diameter growth. The number of competitors to be cut depends on the development phase, the diameter growth wanted, and the thinning cycle. The competitors with the greatest negative impact on the valuable trees should be removed first. In the phase of crown expansion, the crown base should be kept at a fixed height. This requires regular interventions as valuable broadleaved tree species often are getting less competitive with increasing age.

The time of harvest is determined by the time of maximum average value production, the marked situation and other factors. High-quality timber of large dimension consistently realise high prices on the market and demand exceeds supply. The wood of lower quality is of far less value. The answers to the questionnaire 1998–2000 clearly predict an increase of the relevance of veneer wood. Such wood should be straight as well as defect- and branch-free. The minimum bole length should be 2.5 m or even better 6–7 m and the diameter 50–60 cm. The idea of fixed rotation may be abandoned as trees of the same age can grow in groups. Group selection cut may be preferred as larger gaps are needed to fulfil the light requirements of the new generation.

Even so the demand for valuable timber is increasing, and there is a notable interest among forest owners and farmers to grow valuable broadleaved species, on a global scale the relevance of growing valuable broadleaves is still low. However, these species offer an option to produce high value timber in a relatively short time. They may as well fulfil the needs of stakeholders such as small forest owners, farmers and the wood industry. The new options for management include trees on small as well as on large holdings, along roads, in private and public forests of various densities and in mixtures with both broadleaved and coniferous trees. They allow effective production of high-quality hardwood from valuable broadleaved trees with methods adjusted to ownership, tree species, site conditions and type of land and region.

On certain sites or under certain circumstances, valuable broadleaved trees are exposed to various kinds of pests, diseases and other disorders. Potentially damaging factors must be identified before making the decision to plant or cultivate a certain tree species in forests or in plantations. In many cases control possibilities for diseases and pests are limited. However, knowledge of predisposition, genetic qualities of the host tree and factors in disease and pest development can help reduce risk and select suitable tree species for a given site.To avoid major risks plantations should not be very big and mixture with other species in groups or small stands are recommended.

The proposed management options at the same time also increase the supply of non-wood products and services, as well as the diversity of habitats through the varying forest structure and light regimes. Using the ideas of agroforestry in combination with the management of valuable broadleaved trees may create new and innovative management regimes. New ways of multiple land uses may arise, as better habitats offer opportunities for raising livestock, and may also generate jobs and new business ideas, ultimately offering people a possibility to make a living other than from pure farming and forestry. Valuable broadleaves at the forest edge may help to grade the edge from the full forest height to the shrubs by forming naturally shaped, diffuse edges in an irregular and asymmetric way. They can contribute to the uniqueness and beauty of the area in question, taking into consideration the aesthetic, ecological and economic values of the surroundings. As urban forests are becoming increasingly important for people living in cities, cultural values get more attention. Even so broadleaved forests become more popular, species often are not as important to people as the size or structure of the forest.

Valuable broadleaved tree species offer options for increasing ecological, economic and social values and may contribute to sustainability of forestry in Europe and other parts of the world. They may increase the production of high quality timber while maintaining and improving environmental values such as biodiversity, stability and naturalness. However, the high diversity in sites, ownership, economic and socio-cultural conditions in Europe require different strategies adapted to the local needs.

6.3 Future Research Needs and Challenges for Growing Valuable Broadleaved Trees in Europe

Heinrich Spiecker
Institute for Forest Growth, University of Freiburg, Germany

It becomes evident that the information available on growing valuable broadleaves is still rather limited. Demonstration plots representing different management scenarios should be distributed at various locations in Europe on adequate sites in both agricultural landscape and forest rich regions. The unique conditions of each region and landscape type are to be taken into account when determining, where the plots should be situated. The demonstration plots show examples for different management options and visualize effects of spacing on crown architecture, mixture of tree species, plant material and site requirements, pruning and quality control, volume growth control, thinning and final cutting systems and how to optimise the production. They may provide information for forest and farmland owners, private forest owners, forest managers, policy makers, nature park administrators, nature conservation groups, forest visitors and landscape architects. They as well may be used as research objects for researchers.

New tools for inventories able to objectively control the development of wood quality are needed. Indicators for the value relevant wood properties are needed as well. Future management regimes have to take into account the aims and capacity of the forest owners; small forest owners and farmers as well as large forest enterprises. The management regimes should be able to adapt to new socio-economic conditions, as to urbanisation, the demands of the people using urban forests and forest areas close to settlement. Valuable broadleaves may help balancing wood and non-wood production, while maintaining economic attractiveness of forests. The results have to be brought to the target audience in an adequate form.

Appendices

Appendix A: Definition of Valuable Broadleaved Forests in Europe

Sebastian Hein
Institute for Forest Growth, University of Freiburg, Germany

Valuable broadleaved species in Europe: a disambiguation

Among broadleaved trees some species have always attracted extraordinary attention. Outstanding wood properties, aesthetics, a high economic value or simply their rareness have often led to an increased awareness of these species. The term "valuable broadleaved species" reflects the special relevance attributed in forestry to some broadleaved species. This term describes a group of species separate from the normal and ordinary.

Even though "valuable broadleaved species" is a term commonly used in forestry, the use of the term is still diffuse. Some authors have assigned different species to this group (e.g. Bernetti and Padula 1983, Thill and Mathy 1980, Puchert 1974, Koltzenburg 1973), but there is no consistent definition available. Which species really is valuable is often a matter of personal opinion. Indeed, there has always been a consensus in European forestry that such a term is needed: in many countries a concept of "valuable broadleaved species" does exist even though the content of the term is vague.

A survey among experts of valuable broadleaves in Europe has been conducted to better understand the concept of "valuable broadleaved species" to see which species are commonly included, and which silvicultural strategies are currently being applied (Thies and Hein 2000, cf. Chapter 2.2 in this book). In total, experts from 24 countries were questioned. Most of the experts included common ash (*Fraxinus excelsior* L.), sycamore (*Acer pseudoplatanus* L.) and wild cherry (*Prunus avium* L.) in the group of "valuable broadleaved species".

Experts were also asked about the reasons why these species were considered to be "noble" or "valuable". Among various answers most often mentioned were high wood prices, aesthetics of their timber, fast growth, special site requirements or simply rareness and beauty of the tree.

References

Bernetti, G. and Padula, M. 1983. Valuable Broadleaved Species in Our Forests. Monti e Boschi 34: 5–50. (Original in Italian)

Koltzenburg, C.H. 1973. Resources and Use of Valuable Broadleaved Species in Germany – Part I. Harvest and Foreign Trade. Forstarchiv 44: 193–200. (Original in German)

Puchert, H. 1974: Valuable Broadleaved Species – Reflections on Terms and Silviculture.

Allgemeine Forstzeitschrift 29: 977–978. (Original in German)
Thies, M. and Hein, S. 2000. Questionnaire of Experts on the Role of Valuable Broadleaved Species in German Speaking Countries. Sektion Ertragskunde im Deutschen Verband Forstlicher Forschungsanstalten, Kaiserslautern, 05.–07. Juni 2000: 256-269. (Original in German)
Thill, A. and Mathy, P. 1980. Silviculture of Valuable Broadleaved Species in Belgium. Annales de Gembloux 86: 1–32. (Original in French)

Appendix B: Distribution of Valuable Broadleaved Forests in Europe

Sebastian Hein
Institute for Forest Growth, University of Freiburg, Germany

Composition and distribution of valuable broadleaved species in Europe's forests

The proportions of common ash, sycamore and wild cherry in European forests can be estimated by using various criteria: e. g. share in area, volume, increment and timber supply, and even age or diameter class distributions. As the timber or woodland assessments of the UN-ECE/FAO do not offer information on the contribution of single species, especially of minor species, and even less on a country-by-country basis (UN-ECE/FAO 1986, 1992, 1996), results of national inventories were used in this comparison for more detailed information.

Due to the fact that there are different classification systems of an "area covered by forests" or a "forest stand", differences in diameter margins for the calculation of volume and because the rare species are often clustered in groups, inventories have to be carefully interpreted (Päivinen and Köhl 1996). The list of source literature includes the results of the national inventories – inventories on a nationwide scale are generally available and some are even published on the internet – as well as background information from national journals on forestry science. Nevertheless, national inventories or other national surveys and databases can offer insight into the share of valuable broadleaved species in European forests (Table 1) and can facilitate comparisons. Different data sources can lead to inconsistencies, as reported by Schöpfer (1993) in a detailed analysis of the portion of common ash and sycamore in Baden-Württemberg, one of the federal states of Germany. The data presented here had to be modified somewhat to make them more comparable and some comments concerning the source were added where necessary. These comments are noted in an additional column.

For the majority of the countries no species specific statistics are available. The shares of common ash, sycamore and wild cherry are minor in all European countries (Table 1). Where data on a species basis are available, the proportion of valuable broadleaves rarely exceeds 5% (e.g. Great Britain and France). A survey on genetic aspects of forest resources in Europe conducted in 1996 by the Noble Hardwood Network, which included eastern European countries such as Croatia (Gracan 1996), Lithuania (Baliuckas et al. 1996), Latvia (Baumannis et al. 1996), Slovakia (Longauer and Hoffmann 1996) and Slovenia (Pavle et al. 1996), also confirmed the minor importance in area or volume of valuable broadleaves. These values are consistent with the results of the questionnaire on the "Management of Valuable Broadleaved Forests in Europe" (cf. Chapter 2.2 in this book, Thies and Hein 2000). However, it should be noted that on a smaller scale, such as in smaller forest enterprises and under specific site conditions like flood plain forests or plantation forestry, these proportions can increase to 25 % of the total forest area (Thies and Hein 2000).

Table 1. Percentage composition of valuable broadleaved species in selected European countries with a focus on common ash, sycamore and wild cherry.

country	common ash	sycamore	wild cherry	group of species	additional comments
Austria				5.9%*	* other hardwoods; shares expressed in area. (FBVA-AT 1995)
Belgium				11.0%*	* valuable broadleaved forests, including *Quercus, Fagus, Acer, Prunus, Ulmus* stands in which – except *Fagus* and *Quercus* – all tree species in pure or mixed stands take up more than 2/3 of the basal area, not including, e.g.: *Fraxinus* in stands dominated by *Fagus* and *Quercus;* shares expressed in area; data from the Walloon region. (MIN-BE 1997)
Croatia	3.2%*	0.3%*	0.2%*		* incl. *Fraxinus angustifolia, F. excelsior, F. ornus; Acer platanoides, A. pseudopl., A. campestre, A. obtusatum;* shares expressed in volume. (MIN-KR 1996)
Czech Republic				9.8%*	* other broadleaves except: *Quercus, Fagus, Betula;* shares expressed in area. (MIN-CZ 1997)
Denmark	2.6%*	1.9%*			* Acer; shares expressed in area. (Larsen & Johannsen 2002)
Finland				0.0%*	* other broadleaves except *Betula, Populus trem., Alnus;* shares expressed in area of forest land, dominant tree species. (METLA-FI 2002)
France	4.4%*	1.2%*	1.2%*		* shares expressed in volume. (IFN-FR 2003)
Germany				2.9%*	* "other long living broadleaves": *Carpinus, Fraxinus, Acer, Tilia, Ulmus, Prunus, Robinia, Sorbus torm., Castanea;* shares expressed in volume; data from West-Germany. (BMELF-DE 1990)
Great Britain	5.4%*	2.8%*			* shares expressed in area. (FCOM-UK 2002)

Table 1. continued.

country	common ash	sycamore	wild cherry	group of species	additional comments
Italy				10.4%*	* high forest-category; other broadleaves except *Fagus, Quercus;* shares expressed in volume.
				18.7%**	** coppice with standard-category; other broadleaves except *Fagus, Quercus, Carpinus, Castanea;* shares expressed in volume. (ISPER-IT 1985)
Luxemburg	0.2%*	0.2%*			* Fraxinus and Acer from high forests; stand by stand query, additional *Fraxinus* and *Acer* also available in mixed forests; shares expressed in area. (SAV-LU 1994)
Netherlands				5.0%*	* other broadleaves except *Quercus, Fagus, Populus, Salix;* shares expressed in area. (CBS-NL 1985)
Poland		0.1%*			*shares expressed in area; data from state owned forest. (Glaz 1985)
Slovenia	0.5%*	2.4%*	0.3%*		* shares expressed in volume. (ZZG-SJ 2000)
Spain				3.2 %*	* other broadleaves except: *Fagus, Quercus, Castanea, Eucalyptus;* shares expressed in area, by dominant species. (DGCN-ES 1998)
Sweden	0.1%*	0.0%*	0.0%*		* *Fraxinus, Acer, Prunus;* shares expressed in volume; standing volume including dead and wind thrown trees, all land use classes, 1997–2001. (DEP-SE 2003)
Switzerland	2.8%*	2.2%*			**Fraxinus excel./ ornus, Acer;* shares expressed in volume. (WSL-CH 1999)

Regardless of different definitions of mixtures (Köhl and Päivinen 1997), common ash, sycamore and cherry play an important role in mixed forests. Stands with mixed ash and maple are among the most frequent types of mixtures in Denmark, Italy and Great Britain (Bartelink and Olsthoorn 1999). According to these authors mixed stands with common ash and sycamore are even counted among the economically most attractive mixtures in Belgium and Denmark.

References

Baliuckas, V.; Danusevicius, J.; Gabrilavicius, R. 1996. Genetic Resources of Noble Hardwoods and their Conservation in Lithuania, with Special Emphasis on *Fraxinus excelsior* L. and *Quercus robur* L. In: Noble Hardwoods Network: Turok, J.; Eriksson, G.; Kleinschmit, J.; Canger, S. (eds.), Escherode, Germany, 125–132.

Bartelink, H.H. and Olsthoorn, A.F.M. 1999. Mixed Forests in Western Europe. In: Olsthoorn, A. A. et al. (eds.) DLO Inst. for Forestry and Nature Research. Management of Mixed-Species Forest: Silviculture and Economics, Wageningen, IBN Scientific Contributions, 9–15.

Baumannis, I.; Birgelis, J.; Gailis, A. 1996. Genetic Resources of Noble Hardwoods in Latvia and their Conservation. In: Noble Hardwoods Network: Turok, J.; Eriksson, G.; Kleinschmit, J.; Canger, S. (eds.), Escherode, Germany, 133–140.

Bernetti, G. and Padula, M. 1983. Valuable Broadleaved Species in Our Forests. Monti e Boschi 34: 5–50. (Original in Italian)

BMELF 1990. National Forest Inventory 1986–1990 (Volume I+II) Bundesminiserium für Ernährung, Landwirtschaft und Forsten (eds.), Berlin. (Original in German)

CBS-NL 1985. The Forest Inventory of the Netherlands, Volume 1. Centraalbureau voor de Statistiek, hoofdafdeling landboustatistieken Staatsuitgeverij/ CBS- publikaties, Den Haag. (Original in Dutch)

DEP-SE 2003. The Swedish National Forest Inventory 1997–2001, Department of Forest Recource Management and Geomatics, Swedish University of Agricultural Sciences, Umeå, http://www-nfi.slu.se/, (Febr. 4th, 2003).

DGCN-ES 1998. Second National Forest Inventory 1986–1996: Spain. Dirección General de Conservación de la Naturaleza, Ministerio de Medio Ambiente (eds.), Madrid. (Original in Spanish)

WSL-CH 1999. Second Swiss National Forest Inventory: 1993–1995. Eidgenössische Forschungsanstalt für Wald, Schnee und Landschaft (eds.), Bern. (Original in German)

METLA-FI 2002. Finnish Statistical Yearbook of Forestry 2002. Finnish Forest Research Institute, Vammalan Kirjapaino, Vammala, Finland.

FCOM-UK 2002. Forestry Statistics 2002 - A Compendium of Statistics about Woodland, Forestry and Primary Wood Processing in the United Kingdom. Economics and Statistics Unit - Forestry Commission (eds.), Edinburgh.

FBVA-AT 1995. Austrian Forest Inventory, Results 1986/ 1990. Berichte der Forstlichen Bundesversuchsanstalt, Waldforschungszentrum 92, Wien. (Original in German)

Glaz, J. 1985. Forest Resources in Poland. Sylwan, 5/1985: 35–46. (Original in Polish)

Gracan, J. 1996. Present Status of Noble Hardwoods in Croatia. In: Noble Hardwoods Network: Turok, J.; Eriksson, G.; Kleinschmit, J.; Canger, S. (eds.), Escherode, Germany, 45–50.

Harmer, R. and Forrester, M. 1994. Natural Regeneration of Broadleaves in Perspective. Quarterly Journal of Forestry 88: 141–145.

IFN-FR 2003. National Forest Inventory. http://www.ifn.fr; February 10th 2003.

ISPER-IT 1985. National Forest. Instituto Sperimentale per l'Assestamento Forestale e per l'Alpicoltura de Trento (eds.), Arezzo (Original in Italian)
Koltzenburg, C.H. 1973. Resources and Use of Valuable Broadleaved Species in Germany - Part I. Harvest and Foreign Trade. Forstarchiv 44: 193–200. (Original in German)
Köhl, M. and Päivinen, R. 1997. Comparable Data on European Forests. Allgemeine Forstzeitschrift/ Der Wald 23/ 1997: 1266–1267. (Original in German)
Larsen, P.H. and Johannsen, V.K. 2002. Forest and Plantations 2000. Danmarks Statistik, Skov & Landskab, Skov- og Naturstyrelsen, Copenhagen. (Original in Danish)
Longauer, R. and Hoffmann, J. 1996. Noble Hardwoods in Slovakia. In: Noble Hardwoods Network: Turok, J.; Eriksson, G.; Kleinschmit, J.; Canger, S. (eds.), Escherode, Germany, 73–79.
MIN-BE 1997. Forest Inventory of the Wallonian Region in 1996. Ministere de la Région Wallonne, Direction Générale des Ressources Naturelles et de l'Environnement, Direction des Ressources Forestières, Fiche technique n° 9 . (Original in French)
MIN-KR 1996. Forest Inventory of the Croatia, Šumskogospodarska osnova, Zagreb. (Original in Serbocroatian)
MIN-CZ 1997. Forests in the Czech Republic. Czech Ministry of Forests, Praha. (Original in Czech)
Pavle, M.; Smolej, I.; Kraigher, H.; Brus, R. 1996. Noble Broadleaves in Slovenia. In: Noble Hardwoods Network: Turok, J.; Eriksson, G.; Kleinschmit, J.; Canger, S. (eds.), Escherode, Germany, 51-63.
Päivinen, R. and Köhl, M. 1996. EFICS - Towards Improved Forest Information for Europe. In: Nyysönen, A.; Ahti, A. (eds.), EFI Research Paper, 620: 300-305.
Puchert, H. 1974. Valuable Broadleaved Species – Reflections on Terms and Silviculture. Allgemeine Forstzeitschrift 29: 977–978. (Original in German)
Schöpfer, W. 1993. More Options to Analyze Data from the National Forest Inventory. Allgemeine Forstzeitschrift 23/1993: 1186–1192. (Original in German)
SAV-LU 1994. Special Analysis on Valuable Broadleaved Species in Forests of Luxemburg, Service de l'Amenagement des Bois et de l'Economie Forestière, Luxembourg. (Original in French)
Spiecker, H. 2001. Reforestation after Storm with Valuable Broadleaved Species. Freiburger Forstliche Forschung Berichte 25, 89–100. (Original in German)
Thies, M. and Hein, S. 2000. Questionnaire of Experts on the Role of Valuable Broadleaved Species in German Speaking Countries. Sektion Ertragskunde im Deutschen Verband Forstlicher Forschungsanstalten, Kaiserslautern, 05.–07. Juni 2000: 256–269. (Original in German)
Thill, A. and Mathy, P. 1980. Silviculture of Valuable Broadleaved Species in Belgium. Annales de Gembloux, 86: 1–32. (Original in French)
Thoroe, C. and Ollmann, H. 2001. The Future Development of the Woodmarket in Germany, Europe and the World – Potential for Fast Growing Tree Species? Forst und Holz 56: 75–80. (Original in German)
UN-ECE/FAO 1986. European Timber Trends and Prospects to the Year 2000 and Beyond. United Nations Economic Commission for Europe; Food and Agriculture Organization of the United Nations, Timber Section (ed.), Vol 1+ Vol 2, Geneva, Rome, New York.
UN-ECE/FAO 1992. The Forest Resources of the Temperate Zones: The UN-ECE/FAO 1990 Forest Resource Assessment. Report-Nr.: ECE/TIM/62, United Nations Economic Commission for Europe; Food and Agriculture Organization of the United Nations, Timber Section (ed.), Vol 2, Geneva, Rome, New York.
UN-ECE/FAO 1996. European Timber Trends and Prospects: Into the 21 Century. Geneva Timber and Forest Study Papers, United Nations Economic Commission for Europe; Food and Agriculture Organization of the United Nations, Timber Section (eds.), Vol 2, Geneva, Rome, New York.

Weber, N. 2000. NEWFOR - New Forests for Europe: Afforestation at the Turn of the Century. EFI – Proceedings of the Scientific Symposion, Freiburg, Germany.
Weidenbach, P. 1991. The Restoration of Forests Damaged by Storm – The Storms of 1990, Causes and Consequences. Beiträge zur Lebensqualität, Walderhaltung und Umweltschutz, Gesundheit, Wandern und Heimatpflege 33, Siegen, Wilhelm-Münker-Stiftung.
ZZG-SJ 2000. Forest Service of Slovenia: Database on Forest Resources, CDROM, Ljubljana. (Original in Slovenian)

www.ingramcontent.com/pod-product-compliance
Lightning Source LLC
LaVergne TN
LVHW020652300926
849161LV00004B/910
* 9 7 8 9 0 0 4 1 6 7 9 5 7 *